RECORDING NATURAL HISTORY SOUNDS

Richard Margoschis

Paperback ISBN 0900602 24 4
Hardback ISBN 0900602 23 6

Cover Picture from an original by Bill Jackson

Published by Print & Press Services Ltd.
69 Beech Hill, Barnet
Herts. EN4 OSW

Printed by Caligraving Limited, Thetford, Norfolk

CONTENTS

1. *Photo: Jack Skeel.*
The author: with Nagra 1V-S, portable stereo tape-recorder and a pair of Sennheiser 815 gun microphones, in home-made wind shields, tripod mounted.

The Author

Richard Margoschis was interested first in the documentary aspect of tape-recording, but some twenty years ago he started to specialise on recording the sounds of nature; he now has a large collection of wildlife recordings, many in stereo, some of which have been used in films, broadcasting and discs. He has had many successes in both the British and International Amateur Tape Recording Contests, and won the Mammal Class in a Contest organised by the B.B.C. on behalf of the European Broadcasting Union, and judged in seven European Countries, in connection with European Conservation Year in 1970. He now sits on the Organising Committee of the British Amateur Tape Recording Contest.

In 1968 he was instrumental in founding the Wildlife Sound Recording Society with the aim of bringing together all those actively making wildlife recordings and providing a pool of information on equipment and techniques. In the same year he was approached by 3M United Kingdom Ltd., for assistance in running a Wildlife sound recording contest and has since been responsible for doing all the pre-judging in the annual 'SCOTCH CONTEST', and presenting the short-list of entries to the Final Judging Panel.

For some six years he wrote a monthly column in Tape Recording Magazine, and a booklet he wrote on the subject was published some six years ago.

Four years ago Richard Margoschis started recording natural history sounds in stereo and has recently published his first cassette of British Wildlife Habitats in Stereo.

Now able to spend more time on what he still regards as a pastime, he has expanded his lecturing activities as well as his work in the field and the studio.

Preface

It is clear that the majority of people who become interested in recording the sounds of nature already have a considerable knowledge of some branch of natural history, ornithology being the most common, and wish to record the sounds they hear when in the field; there are a few, however, who are primarily interested in sound-recording and have a desire to apply their skills to recording natural sounds. I was one of the latter, and experience over the last twenty years has suggested that it is the more difficult approach to the subject of recording natural history sounds on tape.

I have attempted to write this book in such a way that it might be helpful to either the recordist or the naturalist who has become interested in the combined disciplines; to show something of the difficulties of approaching the subject in the field; of making good recordings of the sounds involved; ways in which these recordings can build up a sound library which can then be used, creatively, to produce recorded programmes.

Over the years I have spent many hours in remote parts of the countryside of Britain, at all times of the day and night, during all seasons of the year. In this I have had assistance from many people, and in many different ways – too numerous to mention, but my thanks go to them all. In particular, I am very conscious of the fact that I should not have done what I have, had it not been for the forebearance of my wife; Vivien has encouraged me throughout, and her understanding of my desire to go out on lone recording expeditions has meant a great deal to me.

Richard Margoschis
Mancetter
Atherstone
Warwickshire

December 1976

The Subject

Simple analysis of the sounds of the countryside will quickly reveal that they are of two distinct types, each of which can be diluted by man made sounds. In a woodland area on a spring morning, animal sounds will be predominant but with them may be such sounds as the wind in the trees, or water trickling along a nearby stream; man made sounds due, more often than not, to various forms of the internal combustion engine, might also be present. These man made sounds are usually looked upon as being unwanted noise, whilst those of the animals, the wind and the water, can collectively be put under the heading of 'natural sounds', but a further separation can be made by grouping animal vocalisations as distinct from other natural sounds. There are however, sounds made by animals, for the purpose of communication, which are not vocalisations. If the sound tells something of nature it should be of interest and, under some circumstances, even man made sounds can be involved.

The majority of animal sounds noticed by the average person originates from birds, but it must not be forgotten that there are other sources too; mammals, large and small, reptiles and insects all produce a multitude of sounds, from the roar of the stag at rut to the high pitched hum of the hover fly. Although some of these sounds are purely mechanical, they have their place in animal communication, a subject which appears to be revealing to researchers more and more remarkable facts.

There are many factors which will have an affect upon the sounds to be heard at any particular location. All animals are adapted to a particular type of habitat, some can accept a fairly wide tolerance and so may be common over a wide geographical area, but to others the conditions are much more specific, and their distribution will be more localised, within a geographical area. It is well known that movements take place with the changing seasons, this applies particularly to birds migrating from one part of the world to another, but it can be more localised. During springtime there is a large increase in the number of species as the summer visitors arrive, and this is the best time of the year from the point of view of birdsong, but as these birds move away with the approach of autumn so their places are taken by winter visitors, not so numerous and not in full song but, nevertheless, producing various contact calls.

Many animals are more vocal at certain times of the year, or produce varying sounds with varying seasons; deer produce much more sound during their rutting season, whilst the fox might be heard during almost any month, but the typical mating call is normally heard only during the mid-winter rut.

As dawn breaks, the birds will be heard to start their song to produce a chorus which rises to a peak during the month of June, they go on singing throughout the day and several, not only the nightingale, can be heard at night; some species, like the nightjar, will start to sing as dusk falls. Many mammals, like fox and badger, are principally nocturnal in their habits and so their vocalisations are most likely to be heard during hours of darkness, whilst deer, and some of the smaller mammals, can be heard by day.

All this adds up to the fact that there are sounds of nature to be recorded throughout the twenty-four hours of each of the three-hundred and sixty-five days of the year – and on the extra day of a leap-year! The material available for recording is endless and it is unlikely that any one person would be able to cover it all in the work of a lifetime. It is most probably true to say that the songs of all the birds which visit the British Isles have been recorded on tape, but it is very doubtful if the complete vocabulary of every species has been dealt with. There is much work still to be done in this direction, to say nothing of the world of insects, and, in any case, there is a great satisfaction in building up a personal library. There is a distinct advantage in the fact that, unlike human vocalisations, works of music and so on, copyright is not involved in recording natural sounds; the copyright in the sound, once on tape, belongs to the person who made or commissioned the recording.

For several reasons, the subject is not an easy one to record on tape. In the first place, because the vocalisations and other sounds are coming from animals that are wild and free, it is not possible to 'direct' the subject as can be done in a recording studio – imagine saying to a bird or a fox, "Please stand a little nearer to my microphone". Everything is under the recording engineer's control in a studio, but when the great open spaces are the recording location, and the subject an animal, the situation is very different and little is under the direct control of the recordist. Because the wanted signal is at relatively low level and there are considerable periods of silence, the problem of unwanted noise is one that becomes very real, and the sounds to be dealt with are not always easy to get on to tape without loosing part of their content or introducing other forms of distortion. Nevertheless, as will be seen in succeeding chapters, there are ways of combating most of the problems.

The Approach

Although by no means the ideal location for natural history sound recording, unless it happens to be situated in the heart of the country, the domestic garden is the first spot in which to hear natural sounds. Within the garden area the sounds are not necessarily confined to birdsongs and calls because, undoubtedly, insects will also be present, and often small mammals can be found. Two distinct possibilities are the garden bird-table and nesting box; a microphone can be placed very close to both these sites and, under such conditions, it is often possible to be in a position to be able to exclude much of the unwanted noise because the microphone is so close to the subject. When birds are using a box for nesting then care must be taken not to disturb them, the microphone must be fixed in a semi-permanent position and not moved until the box is vacated. Even town gardens are suitable for this practice for, providing there is some cover in the form of trees and bushes, a large variety of species can be found in populated areas; animals learn that where man lives there is food to be found and it is not uncommon now to find the fox invading suburban areas.

It is well worth carrying out a detailed reconnoitre of the countryside within a defined radius, selected to bring all points within easy reach. In doing this it is advisable to look for places having a variety of habitats and to mark them, for easy reference, on a one-inch Ordnance map. At the outset, the map will give some guidance of the type of country; look for woodland, pasture and open moorland or mountain country, all of which will produce varying species of both mammal and bird life, and remember to pay particular attention to any inland waters or estuaries where waders and water birds are likely to be found. Even the smallest pond is worth investigating from the natural history sound point of view because if it bears no particular bird life it might well be a breeding ground for frogs and other reptiles.

Consider a location, also, in respect of its suitability for sound recording purposes. An open site will produce very disappointing results if there is a motorway or main line railway within a mile or more, but in a valley, especially if well screened by dense woodland, the level of unwanted noise may be acceptable. Should the valley contain a fast flowing stream a problem as difficult as the motorway can arise because of the sound frequencies involved. There are means of removing some low frequency traffic roar without affecting the required recording, but the more random frequencies contained in the noise produced by a stream are much more difficult to deal with. It should be borne in mind that some locations will be more suitable than others under varying weather conditions. There is little that can be done so far as aircraft noise is concerned except to wait for it to subside.

Acoustics can have a bearing on the quality of recordings

even in open spaces and so it is well worth making test recordings at each location; 'life' can often be added to a recording by an echo, even if only at low level, produced by large groups of trees or the side of a valley.

When all these locations are marked on the map they can be colour coded for visits at specific times of the year, according to the species to be found there. After some experience has been gained at each site, it is worth selecting a limited number of the more rewarding ones for regular visits; in this way there is an opportunity of following a particular species through some cycle in its life whilst, at the same time, a sound picture of the location can be built up covering all seasons of the year and times of the day. To really concentrate on such a project it would not be practical to select more than two locations a year.

If the opportunity to explore further afield comes only with holidays, then time can be well spent beforehand by studying a good map of the holiday area together with any handbooks available; by applying experience gained in selecting local spots, possible areas can be marked for investigation. A quick wander around, upon arrival, will consolidate knowledge gleaned from the map study and confirm the best spots to linger as time permits. To discover any specially good places in the area a word with a local inhabitant, such as the policeman, postman or publican, can be rewarding because any places frequented by naturalists will generally be known to one or another of them.

A point which must receive early investigation is whether or not public access is available to the selected locations, when it is not, then contact must be made with the landowner to obtain his special permission. A polite approach, coupled with an explanation of the work which it is proposed to carry out, normally results in the necessary permission being granted even though conditions might be attached; it should be needless to say that any such conditions must be honoured. Particular care should be taken if it is proposed to visit a location during the hours of darkness; in such event it is always advisable to let the owner know of your intended presence whether or not he has requested you so to do. Landowners who have permitted access to their land naturally expect that the person to whom it is granted will be responsible and follow the code of the countryside. There is also the unwritten law of the naturalist which must be borne in mind at all times – never, wittingly, do anything which will be harmful to, or unduly disturb, an animal; particular attention should be paid to this matter during breeding seasons.

Natural history sound recording is a relatively new pastime which is being taken up by an ever increasing number of people who have come to realise that it can offer the challenges and rewards demanded of a worthwhile hobby; upon their shoulders falls the burden to ensure that it does not fall into disrepute by arousing the contempt of landowners and other naturalists, as would undoubtedly happen if damage to property and disturbance of animals was caused. The unfortunate truth is that, in the past, animals *have* been disturbed, and birds especially by all sorts of people under the guise of studying nature.

The Protection of Birds Act 1967 contains many points to concern the ornithologist, and of direct interest to those seeking only to record birdsong is the fact that the Act makes it an offence for any person wilfully to disturb any wild bird while it is on or near a nest containing eggs or unfledged young, if that bird is included in Schedule 1 of the Act. A very useful booklet "Wild Birds and the Law", which sets out the provisions of the Act, can be obtained from the Royal Society for the Protection of Birds at The Lodge, Sandy, Bedfordshire. Other than this legal restriction, very little stands between the recordist and all the sounds of nature; everything possible must be done, by those concerned, to keep it that way.

All around the world, and certainly throughout the British Isles, there is a great awakening of interest in nature generally, and natural history societies are expanding or being founded nearly everywhere; those interested in the pastime but lacking knowledge of nature would do well to join their ranks. Similarly, the naturalist wishing to progress in the art of recording the sounds he hears has the opportunity of joining a tape recording club and should consider the very specialised Wildlife Sound Recording Society mentioned elsewhere in this book.

The modern battery operated, and therefore completely self-contained, tape-recorder weighing from fifteen pounds down to as little as two or three pounds, is such that it can fairly conveniently be carried over considerable distances. In recent years people have started to carry them around rather in the way that cameras have been carried for many years, and it is certainly not uncommon to see ornithologists using them, especially the light weight cassette type, with the microphone held at arm's length. This might well prove satisfactory when sound recording is looked upon as secondary to ornithology, or any other branch of nature study in which sounds are involved; effectively, the tape-recorder is being used as a note book. The method will very quickly prove unsatisfactory to either naturalist or sound recordist wishing to take up the art of recording nature's sounds seriously.

Before considering the various points to be watched when making a recording it is worth carrying out a simple exercise which can be done in a garden, or where ever a bird can be heard singing and, at this stage, identification of the species is quite unnecessary. Select a suitable spot and wait with the microphone in your hand and the recorder ready to go as soon as a bird comes within twenty or thirty feet, and use full record gain. Do not be discouraged if at first nothing comes your way – it will, eventually! In selecting the spot for this first exercise pay no attention to traffic or any other unwanted noise, in fact its presence can be of assistance in the next exercise.

When you replay this recording you will very soon realise that it is almost impossible to get near enough to a bird to make a good recording, when you are holding the microphone in the hand. Certainly, you should be able to hear the birdsong, and that alone can be quite a thrill if it is a first attempt, but the level of unwanted noise will, almost certainly, be so high as to seriously spoil it as a good recording – to say nothing of the 'blasting' caused by even a light breeze on the face of the microphone.

Whilst waiting at the recording location you might well have noticed that the bird persistently used one or two perches from which to sing, especially if it happens to be the breeding season. Now return to the same location and fix your microphone as near as possible to one of the perches used; three to four feet should be a good starting distance. Run a lead from the microphone and plug it in to the recorder at a point from which you can conveniently watch for the bird to arrive. Commence recording as soon as he starts to sing, using a low microphone gain at first but gradually increasing it until full gain is in use; a quick return of the gain control to zero before each increase will provide easy recognition upon replay. If an extension lead is not available, leave the recorder running near the perch – that will suffice on this occasion, but set the record level at a lower gain than that used for the previous exercise.

Upon replay, the effect of the microphone position will immediately be noticeable, for the voice of the bird will be much stronger and unwanted noise should have faded into the background, even if it has not disappeared completely, and wind 'blasting' will also be considerably reduced because of the lower gain in use. The gradually increasing gain will show the effect of 'under' and 'over modulation', a matter dealt with in a later chapter.

It is always advisable to be prepared to start recording at a moments notice. This applies on such occasions as when walking to a chosen location or even when taking a picnic meal, so often the unexpected occurs or some rarely seen activity is witnessed at close quarters, and failure to record any sounds involved might be to miss the chance of a lifetime. For this sort of work it is an advantage to have the microphone mounted in a parabolic reflector which can be carried in the hand; this piece of equipment is discussed in a later chapter.

By regularly visiting selected locations their potential at any time of the year becomes known, and their use second nature, but if this intimate knowledge is not available then a survey is necessary to determine the best course of action. It might be known that a certain species is present in the habitat, in which case it is necessary only to decide upon the best observation post, but if it is known only that the habitat is right for, and likely to contain, the required species, then there is the added necessity of determining whether or not it is present – this, incidentally, is a time when, if birds are the subject, a good knowledge of birdsong is a very great help.

Song Posts

Most small birds are very territorial in their habits, much of the purpose of their song being to advertise to others of their species the boundaries of the territory they hold, though their voice is, of course, used for communication as well. Song is frequently delivered from one of a number of selected perches and, after an observation session, use can be made of this fact by placing a microphone near a 'song-post'. A couple of chemical retort clamps fixed together can be a very useful tool for this purpose; one clamp holds the microphone and the other is fixed to the branch of a tree or other suitable fixture. The microphone should be set so that the bird, when it sits at its post, will sing to the face of the instrument because if the song happens to come from the rear some quality can be lost. There is nothing to prevent more than one microphone being fixed, each at a different song post, depending upon availability and sufficient leads to run to the observation point, but if this is done the lead to each microphone must be clearly marked to facilitate connection to the microphone at the song post chosen by the bird; a selector box, through which all microphones are connected, can provide the facility of switching from one to the other instead of changing plugs.

An observation post must now be selected from which all

microphones can be seen, and one which will give some cover. As much comfort as is possible under the conditions prevailing is always worth aiming for when a fairly long wait is anticipated; a small fisherman's stool or nothing more than a piece of plastic material can help. Sufficient cover to break up the form of the human body can be obtained by sitting against a tree trunk or near a bush. Again, because of the likelihood of unexpected events, it is advisable, if possible, to have an additional microphone at hand – there is little more exasperating than to be unable to record something happening just in front of you because all available microphones are some distance away. Under these circumstances a microphone in a reflector is a good idea. Such diversions can, however, be dangerous or even disastrous, to the job in hand. When sitting waiting, the time is best spent by observation of any activity, but the main subject must not be forgotten; it often happens that two events occur simultaneously and a very quick decision has to be made as to which is the more important. Experience will allow anticipation to be coupled with observation and so improve the chances of being prepared for the unusual event to happen.

Call Down

Because birds use song to delineate their territory they will invariably react to the sound of another voice of the same species within that area, and this fact is sometimes used to advantage. For instance, if the pre-recorded song of a robin is replayed at the recording location it will very soon 'call down' any robin holding territory there; his first reaction is usually to sing, because his first line of defence is to drive the intruder away by this means. Similarly, if a bird is singing in a half-hearted sort of way he can often be stimulated to a better performance by recording and replaying some phrases of his own song. If this replay is continued it can cause a bird to become extremely agitated, and for this reason the practice **must be used with great care**, especially during the nesting season.

It is sometimes used for purposes other than those connected with sound recording and many ornithologists hold very strong views on the matter; some go so far as to consider that under no circumstances should it be used. **If a sound recordist does, for any reason, employ 'call-down' (or 'play-back'), then he must be sure that he knows what he is doing and bear in mind what has been said about its effect on a bird, and the law relating to wilful disturbance.**

There are circumstances when it can pay dividends, and indeed when it is advisable, to have a microphone in a location for long periods. The situation usually arises from a desire to make a series of recordings covering the progress of an animal; for instance, the changing hunger calls of chicks in the nest from the time they hatch to the time of fledging, and the sound of parents bringing food. A lot too much disturbance would be caused by putting the microphone in place for each recording session, but placing it in position before, or immediately after, hatching **and while both parents are away from the nest**, should not cause disturbance if done quickly. Once in place it must remain there until the young have flown, because any attempt to remove it might cause them to leave the nest before being ready to do so. To protect it, the microphone can be placed in a plastic bag suitably sealed with adhesive tape, care being taken to see that no loose parts of the bag can be blown against the body of the microphone and so cause noise; joints in the cable from the microphone can be similarly treated. If the microphone is mounted in a reflector it will not be necessary to go so near to the nest; this method is very worthy of consideration if circumstances permit its use.

It can not be stressed too strongly that work close to a nest might constitute an infringement of the Protection of Birds Act unless the necessary licence is obtained.

Balance

Considerable care must be taken to obtain a satisfactory

balance between the song, or call, of the species being recorded and any other sounds. This relationship will depend upon the strength of the song it is wished to record, the strength of any song from nearby birds, and the distance of each from the microphone. If the recording is to be of an 'individual species', then the song of that species should be strong and dominant; it must not be mixed up with a lot of other songs, but the presence of other sounds at low level will have the effect of 'placing' the species in a habitat or atmosphere. Going to the other extreme, with no background sounds at all, produces a 'clinical' recording; at times this situation can be desirable and such a recording is particularly good for mixing with others during a copying process.

The positioning of a microphone, whether it be in a reflector or not, is, therefore, of prime importance. The nearer it is to a bird, the more that bird's song will stand out in front of any other song until, if very close, all background will be lost and the recording becomes 'clinical'. It must be remembered that a reflector is very much more directional than any open microphone and so can help in isolating the song of a single bird but not necessarily to the exclusion of all others.

Signal to Ambient Noise Ratio.

It has already been shown that one of the greatest problems facing the outdoor recordist is interference from unwanted sounds, such as traffic. There are, of course, instances when a certain amount of what would otherwise be regarded as unwanted sounds can be tolerated. For instance, a party of swifts screaming as they fly around the houses on a summer evening would not necessarily be ruined by voices, or even the odd motor car, providing these sounds were not dominant.

The 'signal to ambient noise ratio' for any given situation depends entirely upon the distance between the microphone and the sound source, whether it be an open microphone or one mounted in a parabolic reflector; using the same equipment it can be improved only by moving the microphone nearer to the source. Even an adjustment in the record gain, which in effect increases or reduces the sensitivity of the microphone, will make no difference because all signals received will be similarly adjusted. As the distance to the source is reduced, so the sound power received is increased and, unless the distance to an unwanted source is also effectively reduced, the ratio will be improved. To reduce the distance by half is to increase the power fourfold (+6dB), and a further halving will do the same again. This is an important point and it must be considered in conjunction with the desired balance described above.

Other factors, of course, have a bearing on the ratio, the most obvious being to adjust the recording position to place the source of any unwanted noise behind a directional microphone or reflector. However, any movement is likely to cause disturbance resulting in the loss of the subject, if only temporarily, and so it is an insurance to obtain a recording first, and then attempt an improved one by manouvering or waiting for the subject to take up a more favourable position.

Identification

Great care must be taken to ensure that the sound being recorded is that which it is thought to be; if the recordist is not absolutely certain that the sound he hears is being produced by the bird he is watching and recording, then mistakes in identification can easily occur. The situation can arise, for instance, in which a bird is visually identified and thought to be singing when, in fact, the song is coming from a different bird hidden by foliage; this situation is not confined to birds, it can happen with mammals and insects as well. A similar predicament is presented when an unknown sound can be heard but there is no visual evidence as to what is producing it, the answer is to record it first and then investigate to see if the source can be determined.

The golden rule is that unless the recordist is **absolutely** certain on aural identification alone, then identification **must** be supported visually, but if neither aural nor visual identification can be accomplished, then assistance will be required later. Any good recordings in doubt should be filed under a special 'unidentified' heading because it is always possible that later experience, or the aid of an expert, might provide the answer. If an expert is to be consulted on the matter, then he will require to know all the relevant documentation; this is dealt with in a later chapter.

A lot more goes into the making of good natural history sound recordings than has so far been described, but at least, several points have been brought to notice. The choice of location is important from a point of view of both wanted and unwanted sounds; a study of the animals' habits can give guidance on microphone placing; correct recording techniques are necessary and, above all, a very high degree of personal patience is absolutely essential – once it is felt that time has been wasted because of the failure to make a recording during any session, then the operator should start to look for another outlet for his talents.

Birds in flight

The reflector is extremely useful when recording the calls of birds in flight; as a rule the moving subject will be in view and visual aiming of the reflector can be relied upon as it is swung in an arc. If calling birds are wheeling around a restricted area, a better illusion of movement can be obtained by holding the reflector steady in one position, the recorded strength of the signals will then vary as the subjects move on and off beam.

Care must be exercised if a reflector is used in this way because if it is panned across open country to woodland and then, perhaps, a stream, considerable changes in background, or atmosphere, will be noticeable. It is best done by holding the equipment in the hand, either on a short handle or a monopod; it is most unlikely that a pan and tilt head can be used for the purpose without introducing unwanted noise.

Mammals

Much of what has been said about recording birdsong can be applied when the subject is a mammal, but several other factors are pertinent and, on balance, this branch of the work is more difficult. The two main differences are that mammals are not normally as vociferous as birds and many of them are principally nocturnal in habit; in addition, they have a very keen sense of smell as well as sight and hearing. Patience is of great importance whenever nature is being studied but here it is of prime importance.

The hours around dawn and dusk can be very productive but with the nocturnal creatures many hours must be spent in darkness, consequently an intimate knowledge of the terrain is essential. Before any serious recording is contemplated it pays to walk over the area in daylight to look for signs of the species concerned, and a couple of nights listening can give good indications as to the most likely spots at which to wait. The animals tend to become territorial during the rutting season and if seen or heard in one place one night, might well repeat the performance around the same time the next night. Some of them come into voice only at mating time, whilst others can be heard throughout the year, with specific calls during the rut – a good knowledge of their seasonal habits is obviously an asset.

As well as disturbance being caused by movement and noise, there is the added problem of scent causing the animal to become alarmed, and so wind direction must be taken into account; the approach to an area in which an animal is expected to be found must be as quiet as possible and in such a way as to proceed upwind. It is wise to go into the area with plenty of time in hand and then to sit still, to allow the disturbance to subside, and await developments.

Wind direction can often be a problem, especially during calm periods when air movement is slight, and upon its determination depends the direction of the approach. Under such circumstances one of the most reliable guides is to release a small bit of tissue held upwards at arm's length, it is very rare that it will fall straight down to earth.

During the rutting season, calls from animals like fox and deer can go on for quite long periods at a time and, providing they are not too distant, recording can commence as soon as a performance starts. If the calls are rather distant there is always the temptation to attempt to stalk nearer; there is no reason why this should not be done, but it does require a lot of experience and so it is always advisable to make sure of getting a recording of some sort first. This applies particularly to species like deer where the rutting calls are given from a confined area, but when sometimes quite fast movement occurs, as can be the case with the fox for instance, it is always better to wait in the hope that the animal will pass within range – still calling!

A series of calls can start at any time and from any spot, few can be predicted with any degree of certainty, and many of them are of such short duration that the sequence is ended before the recording tape is even started. In order to be certain of recording the whole sequence a rather different technique has to be employed. The only sure way is to select the most likely spot and stay there for long hours with the tape running all the time – not necessarily expensive on tape because it can be used over again but batteries can be a problem; a knowledge that there is a reasonable chance of success is really required before employing the method. The tape should always be allowed to run on after any sequence of calls appears to have ended – they can so easily start again without any warning.

When the calls are expected to come from a confined area, a badger sett for instance, then a number of microphones can be set out; one should be near to the entrance and others at various points along the paths leading away from the abode. The microphone cables must, of course, be brought into a central point and rather than laying them along the ground it is preferable, where, possible, to suspend them over tree branches;

the microphones, too, can hang from a branch instead of being placed on the ground. All this setting up must be done several hours before the recording session to prevent disturbance and to allow scent to disperse; as little disturbance as possible should be made in the process. One of the difficulties that might be experienced in this early setting up is to determine the best place for the observation point, because by the time the session starts the wind might have changed direction; there is little that can be done beyond trusting to a fair share of luck! A microphone in a reflector, and mounted on a tripod should be set up at the observation point, it can be used as a second line of defence if sounds are produced out of range of the open microphones.

It is a strange fact that animals appear to accept the motorcar and show little fear when it is stationary, consequently, on the occasions that it can be driven near enough to the location, it can provide a very good hide and observation post.

Absolute familiarity with the operation of the tape-recorder is essential in order to work efficiently in the dark – and this includes loading a new tape. A thin 'pencil' type flashlamp is very useful in emergency, it can easily be screened to reduce the spread of light and, if necessary, can be held in the mouth to leave both hands free.

Small mammals are even more difficult to find than the larger species and their actions less predictable, nevertheless, the same sort of principles can be applied. The sounds they produce are normally at a very low level and so there is the added difficulty in getting a microphone near enough, and the duration is such that there is little alternative to applying the technique of the continuously running tape. Recordings usually come as a bonus when other recording work is in progress, or as a recording made 'on the spur of the moment' – very rewarding, too!

Because of the difficulties of recording these creatures in the truly wild state, consideration might be given to taking them into temporary captivity for the purpose. A special trap, known as the 'Longworth trap', is used for capturing small mammals alive, but it should be used only by an experienced person because special precautions have to be taken to provide food and warmth to prevent the animal dying if left captured for more than an hour or two.

The animals will frequently produce sounds when handled as they are taken from the trap, but such sounds are generally only alarm calls. In order to obtain the more natural calls it is necessary to keep them in a specially constructed cage, suitably furnished to simulate their natural habitat. This is the sort of work which provides a good opportunity for co-operation between the recordist and a naturalist experienced in keeping animals under controlled conditions.

It is, perhaps, arguable as to whether or not truly wild animals will produce typical sounds when held in captivity.

Insects

The world of the insect is one in which a great deal of work remains to be done from a sound recording point of view; in it are included sounds which go beyond the range of human hearing and special techniques have to be used to record them. However, a huge variety of their sounds are well within audible range and normal recording techniques can be applied though some problems, common to other branches, become more acute. The calls are, on average, at very low level and, even when the microphone is very close, a high amplifier gain is required with the result that ambient noise becomes a problem. Many of the calls are at high frequency and so it is possible, either at the time of recording or during copying, to apply a heavy bass cut and so remove much of the unwanted noise; the technique is described elsewhere. Whilst, in this respect, the high frequency element is advantageous it does present problems in eliminating distortion, and is a matter which really does call for high tape speeds and high quality equipment to obtain really satisfactory results.

In the wild state it is very difficult to get the microphone near enough to the insect without disturbing it. Grasshoppers are probably the first insect to attract the attention of the recordist and, if an attempt is to be made to record in the field, one method of approach is to gradually lower the microphone to a point just above the subject, taking care not to allow any shadow to cause disturbance; with this technique must be coupled a good eyesight and much perseverance. An alternative is to have the microphone mounted in a reflector to eliminate the necessity of going in so close; the directional properties of the reflector can also be of assistance in locating the insect's whereabouts.

Whilst it is generally accepted that natural history sound recordings should be made of animals that are truly wild and free – indeed, it should be stated if this is not so – it is also accepted practice to take insects into temporary captivity for recording purposes. If this is to be done, suitable sized jars should be used for the purpose of collection, and care taken to provide food during the period of captivity; release should eventually be made into the same, or similar, habitat.

For the recording process, the minimum requirement is a cage which can be easily constructed of nylon netting stretched over a light wire frame about 18x12x12 inches. Most insects respond to warmth and so it is an advantage to provide this either from a small lamp in the cage, or by means of a reflector-bulb in a free standing lamp beside the cage; care must be taken not to overheat. If it has been possible to obtain a recording in the field, then the replay of this recording is sometimes an alternative means of stimulation. As is the case with other animals, insects have definite calls to attract a mate and varying 'songs' are used during courtship and mating; specimens of each sex should, therefore, be housed in the same, or adjoining cage, and introduced to each other as required. The microphone should be placed as near as possible to the insect to allow the use of a low record gain with consequent reduction of background noise. Having the microphone on an adjustable mount outside the cage gives a better opportunity of adjusting its position without disturbance of the subject. Even in temporary captivity it might be difficult to eliminate all unwanted noise unless the cage can be taken into a sound recording studio, and if any amount of this fascinating work is to be done it would be well worth considering the construction of a 'mini-studio' in the form of a specially constructed sound proof box of adequate size.

A branch of the insect world producing audible sounds are the wood borers, a number of which produce contact sounds by tapping as well as the sound of actually boring; although these are mechanical sounds they are of interest because of being a form of communication. The sound travels well through the timber and is picked up best if the face of the microphone is in direct contact. If the insect happens to be in wood forming a floor then the tape-recorder should not be placed upon it, otherwise the microphone will undoubtedly pick up vibrations from the machine.

Habitats.

An important aspect of natural history sound recording is to be able to reproduce a sound picture of a habitat, referred to as an 'atmosphere' recording when used to provide a sound backing for a film or radio production. To be complete, such a recording must be capable of indicating all the species present within the habitat at a specified time and so, in theory, it will not be confined to birds or mammals but will include both, together with insects and any natural features like streams; in practice it is virtually impossible to include everything within a reasonable duration and so a compromise is necessary.

Such recordings usually concentrate on birds or mammals and efforts are made to exclude other sounds unless species are present which are specifically connected with them – water birds and the sound of water, for example. Whilst the aim is to include as many species as is practicable, no one species should be predominant, and there should not be so many calls at once as to make the picture incomprehensible. The ideal is to have an overall picture in the background with a series of individual species to the fore, and if these latter species can denote a specific activity, so much the better; it is very rare that this can be obtained in the field and the subject is one which, without doubt, requires considerable experience and expertise. The fact should not be overlooked that any one habitat is capable of providing inumerable sound pictures, each varying with the time of the day and the season of the year.

Because the sounds are continually changing the most satisfactory approach is to make long takes on location and later, in the studio, decide upon the sequences to be kept. A considerable degree of editing might also be done in order to remove unwanted sounds and tidy up the finished product.

Microphone positioning is just as important as with other aspects of recording, for there is little use in simply placing a microphone anywhere; it needs to be placed in a spot near where some activity is likely to occur – where birds are known to gather as the tide comes in, for example. Weather is important, and calm conditions are to be preferred unless special features, such as thunderstorms, are sought. This is not to say that habitat recordings can not be made under windy conditions – they can; the sound of wind, if properly recorded and at the right level, can add another dimension to a recording. Adequate wind shields will prevent blasting on the microphone.

An omni-directional microphone placed on or near the ground and pointing skyward, will collect sound from all directions and so present an overall picture, but if a directional instrument is used then, of course, sounds from one direction will be dominant and the 'width' of the picture recorded will depend upon the characteristic of the microphone. It is possible to increase the area covered by employing several microphones spaced some yards apart. Two microphones of the same impedance can be used on the same recorder input if they are wired in parallel and in phase, but a more satisfactory method, especially with more than two, is to use a mixing unit. The microphones should be placed in position and leads run to a convenient point as far away as possible in order to reduce human interference to a minimum.

Reflectors can be used to assist in making habitat recordings but their use will produce a result which is very different from one made at the same location with an open omni-directional microphone. Because of the directional properties of the reflector the sound will be collected from a very much reduced section of the sphere of sound, and if a bird happens to be on the axis of the reflector his song could be dominant. Under some circumstances the reflector can provide definite benefits; its directional property can be used to reduce the effect of unwanted sound coming from one direction, and the lift in signal strength can make a recording possible where it is not practicable to get a microphone deep enough into the habitat – across marsh or water, for example. Occasionally it might be desirable to emphasise a distant sound because of some peculiarity and so to preserve the 'illusion' of one sound predominating over another when the eye of the observer is attracted to its source; a reflector could be the answer.

Within any defined habitat all the animals go about their business of hunting for food, rearing young and so on, and in the process of doing so they are often dependant upon each other. A type of recording which is even more specialised is one which deals with what is going on in just a very small portion of the habitat; it might deal with the activity of only one species but, on the other hand, several species may be involved. The sounds of a parent, bird or mammal, bringing food to its young is a simple example; a predator suddenly appears and all the other species start to make their alarm calls; or it might be no more than a thrush breaking open the shell of a snail by hitting it on a stone. The first two examples given would be sufficient in themselves, but in the case of the thrush at its anvil the picture would be more complete if there was included just sufficient background to set the activity in the habitat.

Many of these activities are the sort which happen quickly and unexpectedly; it is only the alert recordist who is able to successfully get such sounds on to his tape, but there are others, like young fledglings being fed, which can be anticipated and, when seen happening, can be expected to recur within a reasonable spell of time.

Natural Features.

When recording individual species, or even habitats, attempts are generally made to eliminate any sound from natural features, though there are situations in which it is impossible so to do; the dipper, for instance, is a bird normally found beside streams and so it is almost inevitable that the sound of water will be included in a recording of his song but, because his song is the principal subject, the water usually has to 'look after itself', unless detailed observations have been made and a song post located where a good balance between the two can be assured.

Natural features should not be overlooked; they can produce some very interesting sounds and so provide a subject on their own, to say nothing of the fact that they are often required as atmosphere. Water is frequently involved in one way or another, and wind may also be featured; if in proper balance, they can add much to the information presented by a recording.

Unwanted Sounds.

Setting aside the loud voices, screams and yelling produced by fellow humans, the two main sources of unwanted sounds are the internal combustion engine, in its many forms, and the wind. Briefly, unwanted sound is noise, and it is without doubt one of the greatest problems with which a natural history sound recordist has to cope; it can be more demoralising to the patience than any of the vagaries of wild animals. Perhaps we are fortunate in the British Isles that various Regulations have been introduced in an attempt to control noise but, in the view of the author, it is very doubtful if the level of sound pollution can be reduced, and it will be a considerable achievement if it can be prevented from rising. It is a sad fact that there are now few areas in the United Kingdom where noise is no problem to the recordist; it is a product of the technological age, but if it was not for modern technology we should be without the modern battery operated field tape-recorder with which to enjoy ourselves, and so it is probably as well to take a philosophical view of the whole matter. Anyway, there are things which can be done to reduce the effect and, in some cases, to eliminate it completely.

There are many obvious 'do's and dont's'. Well known beauty spots are an attraction and often provide good habitats but, if they are to be visited, it is asking for trouble to do so at a weekend or holiday period. Generally, the level of noise pollution is lowest during the night and gradually builds up to a peak during the morning; consequently, very early morning is one of the best times for outdoor recording. At the height of the breeding season in particular, birds come into song around dawn and gradually the much talked of 'dawn chorus' builds to a climax, but it is not a good time to record individual species because there is so much song that it is difficult to separate one from the other; half an hour later the situation is much more satisfactory. During the afternoon, birdsong tends to be at a minimum but, if there is any, it affords a good opportunity to isolate a single species. The song tends to build up again during the hour or so before dusk and produces the 'evening chorus', this does not reach the same pitch as that of the dawn but can produce some very rewarding results.

Most of us enjoy the benefits of being able to travel easily from point A to point B along a motorway but the noise of doing so will interfere with making recordings for at least a mile on either side – it is continuous. The noise from a railway is of a more intermittent nature, but a busy line will have the same devastating effect. At a greater distance the noise, which falls by approximately a quarter (6dB) each time the distance is doubled, becomes a low rumble – often referred to as traffic rumble; if it is at a sufficiently low level it might be possible to do something about it later when the recording is being processed. On the other hand, it is also possible to introduce a low frequency filter into the microphone line and deal with the matter 'on the spot'.

Land contours can have a considerable screening effect and attenuation can be greater over a shorter distance if a hill is situated between the sound source and the recording location; a valley can give even better protection, unless serious pollution can enter at the open end. Trees do act as a sound barrier but, to be efficient, they must be in the form of a deep and dense wood, a single line of trees will provide virtually no protection to low frequencies.

Fairly large buildings can sometimes be useful but, because the problem is generally the result of low frequency noise, any protection provided will be found only very close to them – within the sound shadow, in fact. Similar use can be made of large boulders in valleys along which a stream runs; within the sound shadow the high frequency 'hiss' will be considerably reduced but the low frequency roar will be hardly affected.

Barriers that will attenuate sound will often reflect sound also; consequently, high ground which is attenuating a noise source, situated on the opposite side to the recording location, may produce an echo – perhaps to advantage – of a sound produced on the recording side. These are all small points but they are well worth remembering and taking into consideration, especially in respect of locations which are to be regularly visited.

The problem of the unwanted human voice is a very difficult one; it is remarkable how far it can carry under certain conditions and, even when at very low level, is liable to spoil a recording. Obviously, if a large number of people are present it is hopeless to attempt any natural history recording, but like the internal combustion engine, a voice can turn up anywhere at an inopportune moment. Surprisingly enough, people are beginning to recognise the equipment used by recordists –

especially if headphones are in use – and if the disturbance is caused by only two or three people, a finger raised to the lips will often have the desired effect; it is always worth a try if the recording of an unusual subject is being interrupted.

A certain amount of reduction of unwanted sound already recorded on tape can be accomplished by filtering and is dealt with in the section covering processing.

Special Studies

For the real enthusiast there are many fields of study in which the tape-recorder can play an important part. It is probably true to say that the song of most birds has been recorded at some time, and the same goes a long way to the truth about other animal sounds; the world of insects is probably the least investigated. Studies can go far beyond the song of a bird, there are all the various contact and alarm calls that are used. To select two or three species and attempt to collect recordings of a complete vocabulary – and to determine the meaning of the calls – would be a worth while project, and one which could last a number of years.

Regular recordings made at a particular habitat throughout the year can demonstrate the differences of season and time of day and, if continued over a number of years, could be valuable in showing changes taking place.

The ideas are endless when some thought is given to the matter.

Analysis of Animal Sounds

Much research work is going on, in the field of Bio-acoustics, into the complex structure of animal sounds and their use in animal communications. How does one bird differentiate between the call of its mate or parent and that of another, for instance? Complicated and expensive equipment is required for detailed investigation but, as is often the case, some work can be done with little more than a tape-recorder capable of operating at varying speeds.

The speed at which the human ear responds to a sound starting and stopping is slow when compared with insects and other animals, consequently, a sound which appears to a human to be a constant one may, in fact, be a series of notes with very

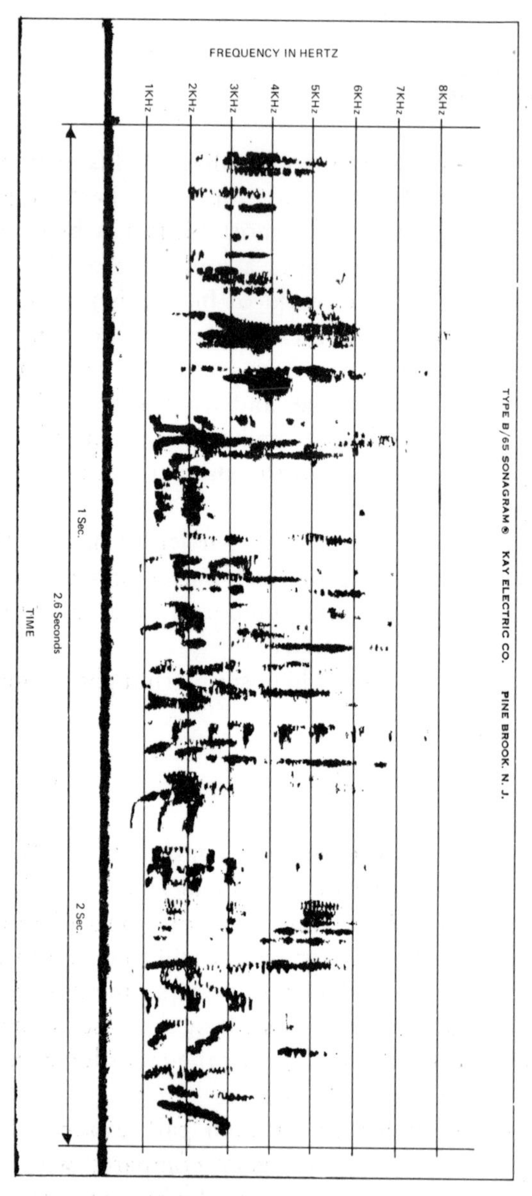

Fig. 1. Sonagram of House Martin song, prepared by D. McNeil and F. Clark from a recording by the Author.

short periods of silence between them.

Replaying a tape at a reduced speed has the effect of extending the duration of any particular note or period of silence; reduction of speed from 15 inches per second to 3¾ i.p.s. will have the effect of multiplying durations by four, and if copying can be done then much greater factors can be achieved. Tonal quality is ruined, of course, but it becomes possible to measure otherwise unheard durations of silence. One method of timing is to measure the length of tape concerned, the time represented can then be calculated.

The sonograph is an instrument which measures the frequencies contained in the signal being fed into it and prints them out on paper, against time, in the form of a sonogram. It provides a very good means of comparing sound sequencies produced by the same animal or by different animals of the same species.

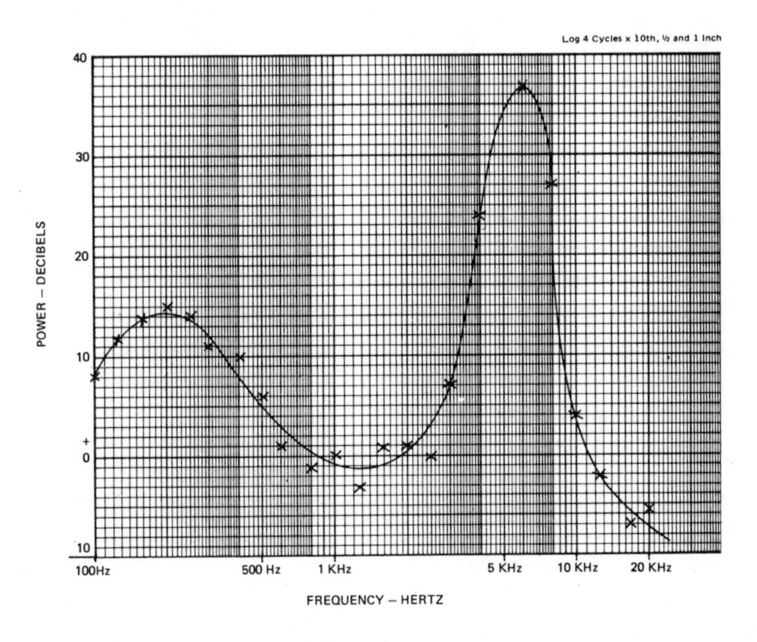

Fig. 2a. Song of Tree Creeper
1/3 octave wave band analysis

The sound can be analysed in a different way by means of the 'octave wave band analyser', an instrument which measures, in decibels, the amplitude of the various frequencies contained and the result can be plotted in graph form.

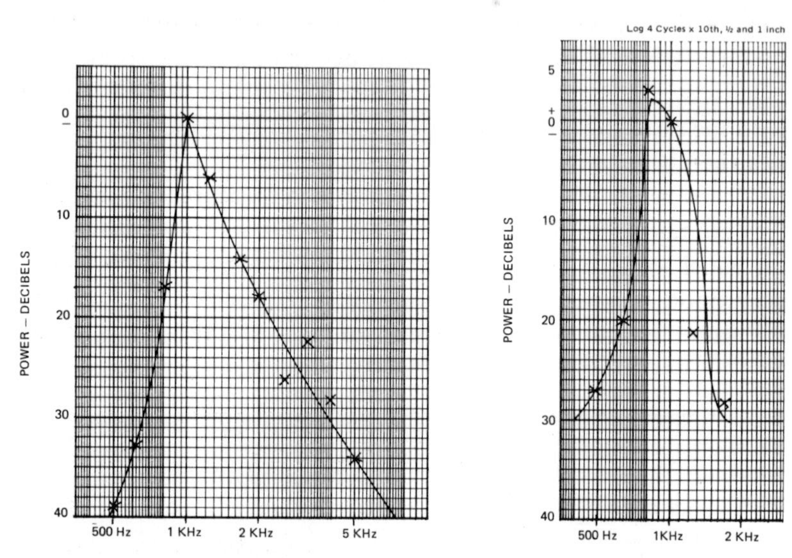

Fig. 2b. Howl of Fox 1/3 Octave wave band analysis

Fig. 2c. Hoot of Tawny Owl 1/3 octave wave band analysis

Ultrasonics

It is generally accepted that the human hearing ranges over a frequency spectrum from 20 Hertz (Hz) to 20 thousand (20KHz) at the higher extremity, and any sound which has a frequency above this is said to be ultrasonic because it is above human hearing. Unfortunately, as a human being ages the efficiency of the ear reduces, so that by middle age the upper limit might well be reduced to something like 15KHz, but is, nevertheless, adequate even then. The fact remains that, whilst human sounds of conversation are of relatively low frequency, there are in the animal world sounds of very high frequency. The most striking are those of the various species of bat which range up to 100KHz. Much exploritory work is going on in the field of bioacoustics.

Although ultrasonic sounds can be recorded on tape they will still not be audible unless manipulated in some way – it can be done. If a recording of a sound having a frequency of 2KHz is replayed at half the recording speed it will be reproduced as a sound having a frequency of 1KHz, but its duration will be doubled; if the replay tape speed is halved again the frequency will be further reduced to 500Hz. By this method, then, sounds which we can not hear are brought within the range of the ear although in modified form, but difficulties arise because, among other things, the very high frequency bias signal used in the recording process can also become audible. Notwithstanding this some very interesting experimental work can be done.

Another method requires special apparatus in addition to the tape recorder. If a pure sound of 1KHz is mixed with a similar one of 3KHz another sound is produced which will have a frequency equal to the difference of the other two – 2KHz in the case quoted.

When bats are in flight they emit sounds of very high frequencies for purposes of navigation and homing on to their prey. An instrument, known as a 'bat detector', has been developed using the difference tone principal to make these calls audible. The instrument is capable of producing a very high frequency signal which can be tuned to different frequencies, and this is mixed with the signal from a special microphone sensitive only to ultrasonic sounds. As the frequency of the sound being emitted by the bat approaches that to which the signal generator is tuned, a difference note is produced on a loudspeaker contained in the instrument. A line out socket allows this signal to be fed directly to a tape-recorder. Different sound patterns are produced by each species.

As its name suggests, the instrument has been developed to help in research into bats, in fact an experienced operator can determine the species of bat present by the sound produced, but it has other possibilities in the field of research into the ultrasonic sounds of animals, and insects in particular.

Under water recording.

There must be many sounds, not yet recorded, made by creatures which live underwater. A hydrophone is sensitive to sound waves in water and is the correct instrument to use for recording such sounds. Ordinary microphones will pick up sounds from underwater but they must, of course, be adequately protected; it should not be difficult to find a suitable rubber sheath for a stick type microphone, the application of which will allow at least part of the instrument to be immersed. Probably one of the greatest problems will be in deciding what creature is making the sound unless individuals are isolated in suitable aquariums.

The Principal of Recording Sound on Tape

It is probably quite true to say that, in practice, it is necessary to know only how to use a tape-recorder; what makes it work is of secondary importance so long as the operator is able to produce from it a satisfactory result. Many people are perfectly happy to work in this way but a basic insight into what happens can be of considerable help, in a number of ways, in making better recordings. Other books, specialising on the subject of sound recording, are available if a detailed knowledge of the mechanical and electronic techniques involved is desired.

Sound is energy which, if produced in air from a point source, radiates equally in all directions in the form of alternating waves of pressure and rarefication; these waves are directly related to amplitude, or loudness, and frequency, or pitch. When such waves impinge on the diaphram of a microphone they cause it to vibrate and so produce in the instrument a minute alternating electrical current which is proportional to both amplitude and frequency. The current is then passed along the wires of the microphone lead to the amplifier of the tape-recorder and here it is 'processed' to make it suitable for passing on to the record head, where a magnetic 'flux' is produced across a minute gap. At this point the electronics of the instrument meet the mechanics which are so arranged as to draw the recording tape across the head at a constant and pre-determined standard speed. The passage of the coating, carried on the tape, through the flux at the gap produces, within that coating, magnetic impulses which are related to the current coming from the amplifier.

Because magnetic tape can be used over and over again, it is necessary that any signal already recorded thereon should be removed before the new signal is applied. This is done by means of the erase-head, also connected to the amplifier, situated in the tape track before the record-head; it is operative only when the machine is recording.

To reproduce the recorded signal the tape must be passed over the head again and so the mechanics provide rewind facilities to allow this to be done. When the machine is then switched to replay, the electronic arrangement of the amplifier is changed and what was a record-head becomes a replay-head; such a head is known as a record/replay-head because of this dual role. Now, as the tape is drawn across the gap the magnetic impulses, contained in the tape's coating, produce within the head an alternating current which is passed on to the amplifier and here the current, which is still related to the sounds originally received by the microphone, is amplified to become of sufficient power to drive the loudspeaker and so reproduce the sound.

What has been described is the most simple form of tape-recorder, and in it certain compromises have to be made because of the dual role of the amplifier and head. A more complicated form of the machine separates the electronics required for the two processes of recording and replay. It requires the addition of a third head in the tape track; the first two heads are used for erase and record in association with a recording amplifier, and the third is a replay head working in association with a separate replay amplifier. Because the two processes are separated it is possible to listen to the recording a fraction of a second after it is made, and a switch is usually incorporated for comparing the recorded signal with the pre-recorded signal – this is known as A/B, or 'off tape' monitoring. An added advantage of the system is that both record and replay heads can be specifically designed for the one job.

In the early days of tape-recording, only one 'track', taking up the full width of the tape, was laid. The system is still used on professional machines and is called 'full track recording'. As techniques improved, and for reasons of economy, machines for 'domestic' use started laying two tracks on the tape. Track

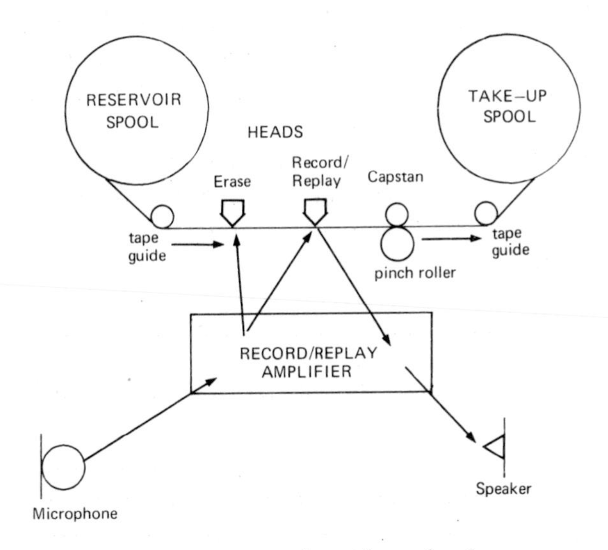

Fig. 3a. Tape recorder with two head system.

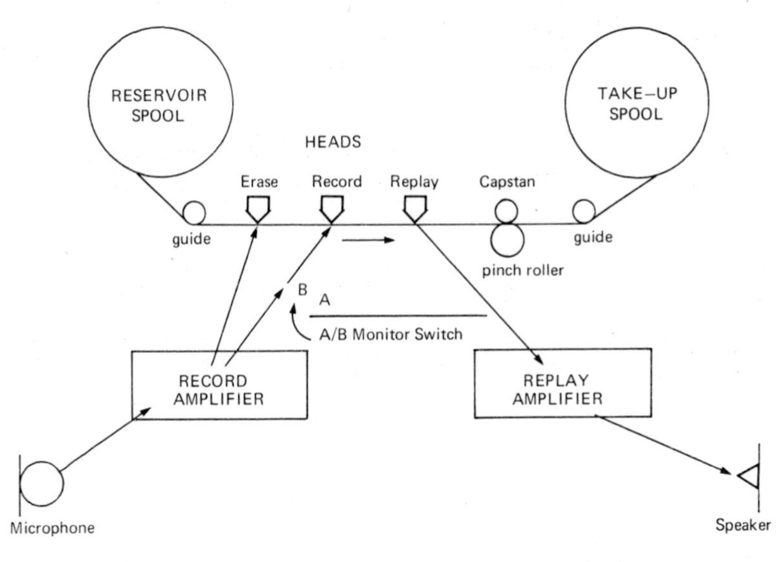

Fig. 3b Tape recorder with three head system allowing 'off-tape', or A/B monitoring.

one is laid on the first pass and track two, after the spools have been turned over, on the second pass; it is known as 'half track recording'. Of more recent years, still further economy has been made by using heads capable of laying four tracks, two in each direction, and known as 'quarter track recording'. Standard operation requires that the tape should pass along the tape track from left to right and that in half-track work the recording is made on the top track.

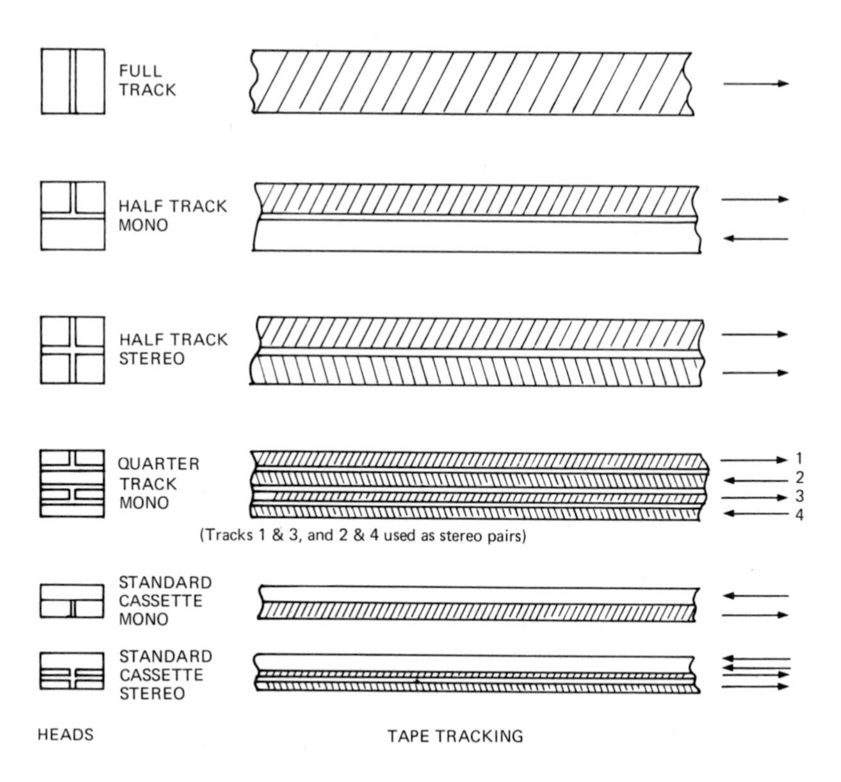

Fig. 4. Tape tracking.

Standards also exist in respect of the speed at which the tape is drawn past the recording heads and, like track width, have tended to be reduced over the years, principally for reasons of economy. The basic standard, used by professional equipment, is 15 inches per second but, in the very early days, this was reduced to 7.5 inches per second for domestic equipment. Further reductions have resulted in tape speeds of 3¾, 1 7/8 and even 15/16 inches per second being used; it will be noted that each reduced speed is one-half of the previous one. Many machines are capable of running at two or more of the standard tape speeds.

Both mechanics and electronics will have a bearing on the quality achieved and, from the point of view of the electronics, tape speed and track width are two important factors. The wave length of a sound having a frequency of 10KHz (ten-thousand Hertz, or cycles per second) is only one-tenth of that having a frequency of 1KHz, and the magnetic impulses produced on the tape will be in the same ratio. If the tape speed is reduced by one-half, then the length of the impulse on it is also halved. As the process of reducing tape speed is continued, a point is reached at which the length of the impulse produced by the higher frequency is so small that the replay head gap is physically incapable of reproducing it at the correct level. When this happens the high frequency response suffers and there is a reduction in the 'brightness' of the recording when reproduced, even though electronic compensation can be included. For high quality recording to be achieved all sounds, from very low – 40 Hz – to very high frequencies – 20KHz – should be recorded and reproduced within very close tolerances.

When a full track signal is recorded, then the replay amplifier is able to make use of the strongest signal which can be presented to it, but if it is reduced to half-track the signal available is reduced by rather more than one-half because a space must be left between the two tracks, and the process is repeated when the track is further reduced to quarter-track. With each reduction the amplifier has to work harder to reproduce the recording and so the 'signal-to-noise' ratio is impared; this is the ratio between the wanted signal and the unwanted noise produced by the electronics of the system.

On the mechanical side, short and long term speed stability are all important. Rapid variations in the speed of the tape passing the head will result in 'flutter', whilst longer term variations produce 'wow' – features which sound exactly as described. An incorrect tape speed – that is, a variation from a standard – will affect the pitch of the reproduced signal; halving the tape speed, on replay, brings the pitch down by one octave.

Making a Sound Recording.

The more complex the tape-recorder in use the more care that is required to ensure that all the controls on the machine are correctly set for a recording to be made. With the majority of machines, all that is required is to see that the microphone lead is correctly plugged in and then either turn a switch to the 'record' position or depress the appropriate key. Because the recording process ruins any previous recording on the tape, most machines require the simultaneous operation of two controls in an attempt to prevent the accidental erasure of an important recording. There must be few recordists who, at some time in their career, have not committed this cardinal sin! It may also be necessary to operate a switch to select the microphone or line input, as required.

However, the very first operation is to load the machine with tape; very simple in the case of cassette recorders but not always so easy when dealing with open spools, especially under adverse outdoor conditions and, perhaps, in the dark. It is practice that makes perfect and irons out any snags so that loading can be effected quickly.

Once the tape is running, attention must be given to the 'record level indicator' and the 'gain control'. The majority of field tape-recorders, operating on a battery power supply, now use a meter as the level indicator but there may still be a few old ones which have a 'magic-eye' instead. The strength of the signal coming from a microphone will vary according to, among other things, the amplitude of the sound and the distance it is from the microphone; it is therefore necessary, at some point between the input to the amplifier and the record-head, to adjust the signal to a level which can be satisfactorily handled by the tape. If the signal being fed to the head is too strong then the tape will be 'overmodulated', and the result will be a distorted recording; on the other hand, should the signal be too weak, the resulting recording will be 'under modulated', and at such a low level that upon replay it will be necessary to turn the 'volume' up so much that the unwanted noise inherent in the amplifier will be heard in the background, resulting in a poor 'signal to noise' ratio.

The aim is to present to the tape a signal which is within the required limits as indicated by the record level meter, and the necessary adjustments to achieve this are made with the record level gain control. The scale on the meter is normally marked with a point at which a peak signal can be recorded but beyond which distortion will commence.

The proper use of the gain control in conjunction with the meter should be thoroughly understood and mastered before attempting to record any natural history sounds. The recordist's

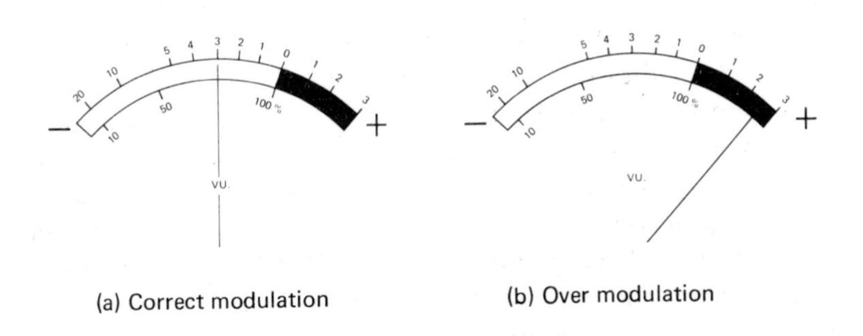

Fig. 5. The scale of a VU meter showing (a) correct, and (b) over-modulation, but see text regarding peaking transients in bird-song.

own voice is a suitable subject on which to practice in the first place, and a great deal can be learnt by replaying recordings made at different levels and distances from the microphone.

When recording a human voice in a studio, or even in a living-room, it is possible to make a number of adjustments in order to achieve the best results. The microphone can be placed in different parts of the room to produce different acoustic effects, the subject can be placed nearer or further away from the microphone and, above all, a test run can be made for check purposes. Once out in the field, with the great open spaces as a studio and a subject which can not be 'directed', the recordist will soon appreciate that nearly all control of the matter has been lost and must be replaced by knowledge of the subject, anticipation and, most important of all – PATIENCE.

It is of prime importance that early experience be obtained in determining the amount of gain which can be used for birdsong without causing distortion. Experiments with the voice should have given sufficient information on which to base the first attempts, but it will be found that with some birdsong – like that of the robin and wren – distortion will be evident when the gain was insufficient to move the meter needle up to the zero point. Unlike the human voice, the power in birdsong is frequently of such a transient nature that the VU (volume unit) meter provided on the majority of tape-recorders is unable to show it. In addition, the meter scale is not linear, and so a small movement of the needle at the bottom end is the equivalent of a much greater movement around the zero mark. These two features, together, mean that the readings of a VU meter can be misleading, in connection with birdsong, until thoroughly understood. The peak programme meter (P.P.M.) has a very much faster action and can read short peaks but, unfortunately, because of its complexity and cost is found on very few portable tape-recorders, and not on many studio machines in the amateur range. Small red lights which flash on when a transient 'peaks', are now appearing on the VU meters of mains operated machines and could be a help if they appear on battery operated machines as well.

The song-post method, as described earlier, is very convenient for gaining experience, either with an open microphone or with a parabolic reflector because, if a good song-post has been selected, the bird should sing from it for periods of sufficient length to allow recordings to be made at different gain settings, and the method allows at least one of the variables to be reasonably constant – the distance between subject and microphone.

The 'dynamic range' of the sounds to be recorded – that is the difference between the levels of the lowest and highest sounds – is likely to be much greater than the recording equipment can handle, with the result that, for any particular gain setting, the lowest sound can be lost in the system noise of the equipment (under modulation) and the highest sound distorted (over modulation). In a studio, arrangements can be made to carefully alter the gain to keep the recorded signal within these parameters but, once again, the situation is different in the field because the recordist will have no warning of a sudden loud burst of song, and any sudden change in record gain to compensate, whilst possibly preventing overmodulation, will result in an alteration of the background level spoiling the recording. In practice, if a subject moves further from, or nearer to, the microphone so much as to alter the required record gain, then the recording should be ended and another started, if conditions are still suitable.

Other items which will affect the strength of the signal which can be applied before distortion from overmodulation occurs, irrespective of the recorder, are the type of tape in use, track width and tape speed; high quality tape and the highest possible tape speed, if more than one is available, will give best results.

When the recording is being made it should be monitored on a small ear-piece or a set of headphones, and in connection with this the difference between pre-tape and off-tape monitoring must be fully appreciated. Most tape-recorders provide a socket from which the signal can be monitored as the recording is being made but, in the case of machines using the two head system, the socket is connected to the amplifier at a point *before* the signal has reached the record-head; consequently, what is heard will be that which the microphone is accepting and will be no indication of the quality of the recording nor, in fact, that a recording has been made because the signal will still be heard even if the tape is not running. The situation is very different on the more expensive type of machine employing a three head system and providing A/B, off-tape, monitoring. A two position switch is provided which will present one of two different signals to the monitor point; in the A position the signal to be heard comes direct from the microphone as in the two head system, but in position B the signal is being taken from the third head and is, therefore, a *reproduction* of the actual recording made a fraction of a second before. Listening to the B signal on good quality headphones will give an indication of the quality of the recording, and by the flick of the switch to the A position it can be compared with the pre-recorded signal; if a signal is present on B monitor then the tape must be running.

Mention must be made regarding the record gain to be used in connection with habitat recordings. The dynamic range in this type of work is very considerable; some, which are well worth making, will contain very little sound and what there is will be at a very low level, whilst others, where the area concerned is more limited and the subjects more concentrated, will be at much higher level. Until experience is gained, a good guide is to start with the gain control set at the level normally used when recording birdsong at average distance.

At the conclusion of a recording turn the record gain down to zero and let the tape run for a few seconds before switching the machine off. This way there will be a gap to facilitate the separation of one recording from another. After the machine is stopped it is as well to return the gain to the average position so that it is ready for the next recording. The importance of documentation is dealt with in another chapter but it is worth mentioning here that a spoken description, recorded immediately following a recording, can be of great assistance later when entering up details in a filing system.

So far reference has been made only to 'live' recording – that is when a microphone is providing the sound source – but recordings can also be made from a previously recorded source. The process of making such a recording is, in fact, copying or 'dubbing', and for this purpose the previously mentioned 'line' input to the recorder is used. If the art of natural history sound recording is to be used to its full extent, then this work must be contemplated; it obviously requires a minimum of two tape-recorders, and other equipment becomes involved. The subject is dealt with separately under the collective heading of Processing.

Microphones

When a tape-recorder is purchased a microphone is often included with the accessories and it is safe to assume that the instrument will suit the recorder, but this does not necessarily mean that the microphone can give the quality that the machine is capable of handling. It is frequently the case that improved results can be obtained by using a better quality microphone, indeed, the more expensive tape-recorders do not include for a microphone in the purchase price because it is considered, quite correctly, that the user will wish to use the type of microphone best suited to the work he intends to do. This fact must not be lost sight of when building up available equipment; one good tape-recorder deserves several good microphones and so some consideration must be given to the matter.

From a constructional point of view there are four main types of microphone; two of them are not suited to the work. The crystal microphone is essentially high impedance and suffers the disadvantage of high impedance microphones described later whilst the ribbon microphone, which has a figure of eight pattern, is so susceptible to wind that it can be damaged by blowing upon its face, and so is not suitable for outdoor work.

The moving coil is the most popular type; it is robust and can be obtained as omni or unidirectional in any impedance. Weight and size, which can be important features when the microphone is used in a parabolic reflector, vary considerably. It can produce very good quality.

The fourth type is the condenser microphone; generally at medium impedance, it is available with either omni or uni directional patterns. It is capable of very high quality and has the advantage of presenting a high level signal to the tape-recorder. They are generally more expensive than the moving coil type and have the disadvantage that they require a power supply to operate. A separate power pack is necessary unless a suitable supply is available from another source. A few battery operated tape-recorders are now capable of giving the required supply and there are versions of the condenser microphone which contain their own power pack; either of these arrangements has the advantage of eliminating the necessity of carrying an extra piece of equipment.

The electret microphone can be described, simply, as a modified version of the condenser and, because it does not require so much power, normally contains a small battery within its case.

The gun-microphone is a specialized form of directional instrument having a very tight cardioid polar response, a physical feature being its extended length – up to twenty-four inches. It is very suitable for natural history work but, whilst it has a better range than an ordinary directional microphone, it does need to be rather nearer to the sound source than a good dynamic microphone in a reflector. Directional properties are hardly as good as a reflector, and so it will not produce such good separation of the songs of two birds which are singing relatively near to each other. On the other hand, there is no loss of low frequencies to the front as is the case with a reflector, and there is very good rejection of sounds from side and rear. It is rather lighter than most reflectors and not so bulky to carry; it can be used in the hand or mounted on a tripod. Signal strength varies considerably, from the dynamic type to the very expensive R.F. condenser type produced by one manufacturer; they are also available as condensers and electrets. Undoubtedly, the various versions of the gun microphone are becoming very popular with wildlife sound recordists.

Modern condenser and electret microphones have their own special pre-amplifier built into the case of the instrument and several offer alternative capsules, or heads, thus allowing the basic pre-amplifier to be used with either omni-direction, unidirectional or gun characteristics. In addition, some versions of the pre-amplifier have built in variable bass-cut filters which are very useful in reducing handling noise and wind rumble.

Radio microphones are available but have a rather limited place in recording natural sounds. At first thought it might appear a good idea to be able to place a microphone at a location and then receive its signals without a connecting lead but, on further thought, it is clear that at least two journeys must be made to place it and retrieve it, at which time it should not be difficult to lay and recover a cable. However, should the laying of a cable be difficult, across a strip of water for example, then the instrument might be given consideration; but they are expensive things to leave about! Because it includes a radio transmitter, the radio microphone comes within the definition of wireless telegraphy and it is an offence under the Wireless Telegraphy Act of 1949 to use one except under, and in accordance with, a licence granted by the Post Office.

All the types of microphone described have three features in common – sensitivity, impedance and polar response. Each can vary.

The sensitivity of a microphone, which is specified in several different ways, determines the strength of the signal being fed to the amplifier of the tape-recorder and will have a bearing on the signal to noise ratio because, with a low level signal, the amplifier gain must be increased to obtain correct recording level, and amplifier noise will become correspondingly more apparent. Natural sounds are of very low power when compared with musical instruments and so this can be a very important feature, but increased sensitivity will not improve the signal to ambient noise ratio.

Impedance is measured in ohms (the sign for which is Ω) and relates to circuits of alternating current. the impedance of a microphone will be given as either low (30-50), medium (200-600) or high (10K+). The microphone input to a tape-recorder can, likewise, vary, and it is therefore of prime importance, when purchasing a microphone, to ensure that its impedance will correctly match the input to the tape-recorder. A high impedance input should receive the signal from a high impedance microphone, if the signal comes from a low impedance instrument there will be a loss in signal strength; if a low impedance input receives a signal direct from a microphone of higher impedance the resultant recording is likely to be distorted. A high impedance microphone can not be used on a long lead without signal loss, particularly in the high frequencies, consequently, low and medium impedances are required.

It is not a simple matter to match high impedance down to low but the reverse can easily be accomplished by the use of a matching transformer inserted in the microphone line, and in this way a low impedance microphone can be matched to either medium or high inputs. If recorders having inputs of different impedances are in use, then a matching transformer might enable a microphone to used with either machine. Fortunately most modern machines, both battery and mains operated, employ medium impedance microphone inputs.

The polar response of a microphone indicates the sensitivity of the instrument to sound coming from different directions.

If it is 'omni-directional', then it will respond equally to a sound coming from the front, sides or rear, but there is likely to be some reduction in sensitivity to high frequency sounds coming from the rear. The 'uni-directional', or 'cardioid', microphone is most sensitive to sounds arriving at the front and becomes gradually less sensitive as the sound source moves around to the side and is least sensitive to sound arriving at the rear. A third type of response is that known as 'figure of eight', when the instrument is equally sensitive to sounds arriving at front and rear but has reduced sensitivity to sounds at either side. Each type has its special uses but omni and uni-directional microphones are the ones most widely used in natural history sound recording. The uni-directional will improve the signal to ambient noise ratio, especially when a particular unwanted sound source is present and provided the microphone can be arranged to face away from it.

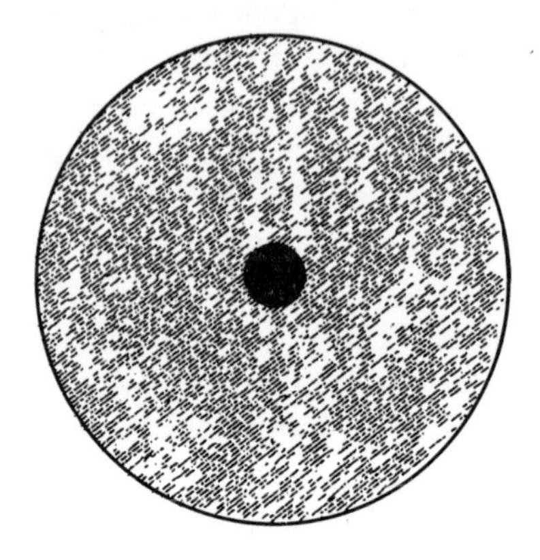

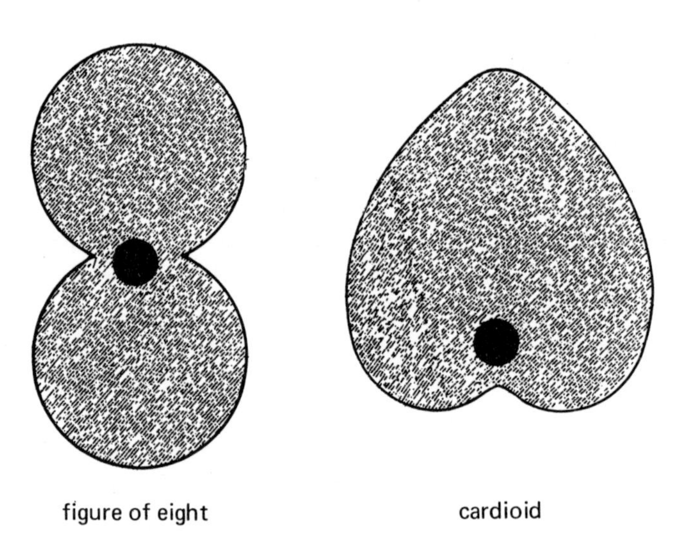

Fig. 6. *The shaded area shows the direction from which strongest signals are received by the three main types of microphone; but note that the precise polar pattern will vary with sound frequency.*

Obviously, there are many points to consider when purchasing a microphone and cost is bound to be included because of the wide price range; a good quality uni-directional moving coil is as good a choice as any to start with.

The Parabolic Reflector

The purpose of a parabolic reflector is to concentrate sound upon a small point in order that a microphone placed at that point can have the benefit of an increased sound power to operate it. Ideally, the resulting concentration should operate equally over the whole frequency range of the sounds to be recorded. In practice, whilst the device does considerably increase signal strength, and makes the microphone highly directional, it does have some side effects. Nevertheless, it is without doubt a very important tool to the natural history sound recordist and, providing it is used correctly, its benefits can far outweigh its deficiencies.

The reflector is constructed to a very precise parabolic curve which is given by the formula:–

$$y^2 = 4\,ax \quad \text{where,}$$

y = distance along the vertical axis
x = distance along the horizontal axis
a = distance of the focal point from where the axes meet.

Consequently, the dish is such that the distance from any point on its plane, to the surface and back to the focal point is constant, and so any sound travelling on a course which is parallel to the axis will arrive at the focal point at the same time irrespective of where it strikes the surface of the reflector. However, the range of sound frequences so reflected will be determined by the diameter of the reflector. As the frequency of sound is reduced so its wavelength increases, and as the wavelength approaches the diameter of the reflector the sound bends around the obstacle and the amount of reflection, and therefore the lift given, is reduced. Sound at a frequency of 500 Hertz has a wavelength of approximately 24 inches, and so a reflector of that diameter will concentrate only those sounds which have a frequency greater than 500 Hertz, and a wavelength of less than 24 inches. There is not, however, a clean cut off at this

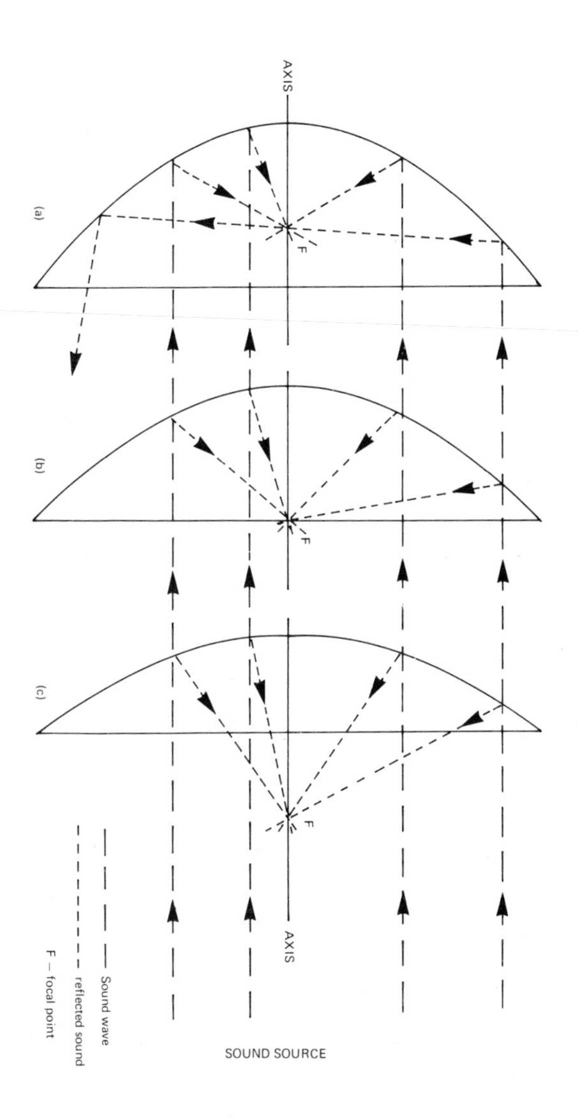

Fig. 7. Diagrams of three reflectors having different parabolic curves. All have diameter of 20 units with focal lengths of:–
(a) 4 units – focal point inside dish
(b) 5 units – focal point on plane of dish
(c) 7 units – focal point outside dish

frequency because, from what has been said before, the efficiency gradually reduces.

The two variables in the construction of a reflector are its diameter and its focal length; if the diameter remains constant the depth of the dish will increase as the focal length is reduced.

The reflector becomes more difficult to handle as focal length is decreased or diameter increased. It is, from this point of view, fortunate that the majority of birdsong contains little sound below 500 Hz and so the fact that a reflector starts to cut off slightly above this point is not too significant; indeed, it might even be a help in reducing unwanted noise such as traffic roar. On the other hand, a habitat recording might sound a little 'thin' if lacking these lower frequencies. The main advantages of the reflector are the greatly increased range over which a satisfactory recording can be made and the fact that, because of its directional properties, an individual call can be singled out from others. It is very difficult to give any estimate of the greatest distance over which a recording can be made because so much depends upon the strength of the call involved and the conditions appertaining at the time, but a medium sized reflector should present a reasonable signal of the average birdsong up to a range of thirty yards, probably more if conditions are ideal – that is, if ambient noise is at a minimum. Even so, the best of recordings could hardly be expected because the record amplifier gain would be high; distance between sound source and microphone still holds good when a reflector is in use.

Reflectors having a diameter of less than about eighteen inches are too small, except that they can be useful for recording the sound of insects which are usually at high frequency; above twenty-four inches in diameter they become difficult to carry and need to be mounted on a tripod when in use. A very convenient size is a diameter of twenty inches with a focal length of either four or five inches; the four inch one is slightly more bulky to carry but, as the focal point is within the dish, it gives slightly more protection from wind than the five inch which has its focal point on the plane of the dish. The latter has considerable advantage in that it is very easy to see that the microphone is correctly placed on the focal point, simply by sighting across the plane of the reflector.

Larger reflectors, thirty six or forty eight inches in diameter, give extra gain on signal strength and so have a greater range as well as being capable of handling lower frequencies; a diameter of about eight feet is required to satisfactorily reproduce a man's voice. Reflectors of twenty inch diameter and five inch focal length are very popular among natural history sound recordists.

A criticism which has been aimed at the parabolic reflector is that it makes the recordings sound 'tinny' or 'bottled up'; the view was held particularly in respect of metal reflectors, glass fibre ones appear to have become the more popular. The use of glass fibre makes it a simple matter to produce either a fairly flimsy and light dish or a stronger but heavy one, in other respects it seems that the performance is virtually the same as metal.

Any piece of equipment, be it electronic or mechanical, can introduce some degree of distortion to the signal and, in the final analysis, it is for the individual worker to decide whether or not the advantages he can gain by using a reflector outweigh some slight variation in the sound. The answer can be obtained, perhaps, only by trial, but it might be worth noting – for those who might be doubtful – that after some one-hundred and fifty people had listened to the song of the same bird recorded with and without a reflector, about half were correct in identifying which was which.

The manufacture of a metal reflector requires considerable skill, for it must either be spun on a lathe or beaten out by hand. Glass fibre is not so difficult to handle and would certainly be the choice for a 'do it yourself' job; probably the most difficult operation is to produce the former on which the glass is laid. Suitable reflectors, both metal and glass fibre, are obtainable but the sources of supply are somewhat limited.

Handling the Reflector

A glass fibre reflector of twenty inches in diameter is sufficiently light for the average person to be able to hold it in the hand whilst recording. provided the microphone mounted in it is a reasonably light one. The position and design of the handle is important; it must be comfortable to the grip and set in a position which will give to the reflector a reasonable balance when the microphone is in position. One of the greatest problems when used in this way is the transmission of handling noise which will be recorded on the tape as a rumble, and so easily spoil an otherwise good recording. A number of precautions can be taken to reduce it as far as possible.

The microphone lead must not be allowed to flap about, it should be brought to a clip on the inner surface of the reflector and near the rim, from this it is taken to another clip on the back of the dish and through which it is made to pass upwards so that it will then fall away to the recorder and be kept clear of the hand. Even variations in the tension of the fingers around the handle will produce rumble, and an arthritic wrist can produce unbelievable noise if moved about! The grip on the handle must, therefore, be firm but steady and movement of the wrist restricted or, if possible, eliminated; the forearm should be horizontal and the elbow pressed closely in to the side, any necessary redirection of the reflector being done by a movement of the whole of the upper part of the body.

When the normal short handle is in use the time over which a reflector can be satisfactorily held, without producing noise, is limited because all the weight is taken through the wrist to the forearm. The problem can be alleviated by the use of a monopod which is really nothing more than the normal handle extended to such a length that it can be conveniently rested on the ground whilst being held more loosely in the hand. Although this removes the strain it does not completely remove the problem of handling noise. If, for instance, the monopod is resting on a gritty surface the slightest movement will transmit noise; the answer is to rest it on the toe of the operator's shoe when it can be turned without trouble. The arrangement does not prevent the reflector from being held in the normal way when it is desired to pan across while following a bird in flight, in fact the extra weight often balances the reflector when the hand is the fulcrum. Telescopic monopods, as used with cameras, are very suitable but an ordinary broomstick, with suitable fitting, is perfectly satisfactory.

With a monopod the operator can cope with quite long periods of waiting and recording, but it is not possible to leave the reflector set up in any selected position, for this, and for really long periods at one spot, the answer is a tripod. After setting up the reflector, and directing it at the desired spot, a lead can be run to a distant observation point at which the recorder can be operated without disturbing the subject. A pan and tilt head mounted on the tripod greatly simplifies the job of directing the reflector correctly but it is by no means essential. From the point of view of mobility a tripod is the most cumbersome of the mounts described but it is often worth the extra effort to carry it because, with the reflector completely supported, both hands are left free to operate the recorder and to use binoculars for observation purposes. Some very light tripods are available at camera shops, they do not need to be fully extended when in use and so are adequate for a twenty four inch reflector.

A minor point, but one which becomes significant if each method described is to be used, is the way in which the handle is fixed to the dish. The fixing point can take the form of a lug which is drilled and tapped a suitable thread to accept either handle, monopod or tripod, but the lug must be firmly fixed because it takes a considerable strain and any movement of it in relation to the dish will produce noise.

Type of Microphone

If an omni-directional microphone is used with a reflector it becomes directional to the frequencies which the dish can reflect but it will also respond to low frequencies which receive no gain from the reflector. The substitution of a uni-directional microphone will result in the loss of these low frequencies because of the very much reduced response to sounds arriving at its rear. Although the equipment reflects sound it is not an efficient sound barrier and so sounds arriving at the rear of the reflector will pass through its substance to the face of the microphone and be recorded, but they will be at a much reduced level because of not having the beneficial effect of being reflected to a focal point.

There is likely to be a dip in the overall response when an omni-directional microphone is used because of sounds of certain frequencies arriving from its rear at the same time as the reflected ones arrive at the front, and so cause cancellation, the frequency at which this occurs varies slightly with differing focal lengths. This cancellation will not occur to the same degree when a uni-directional microphone is in use. It is not a serious loss but it could affect the results of any scientific investigations for which the recording was being used. If a uni-directional microphone is used in a large diameter and short focal length reflector, then some gain might be lost because sound reflected from the area near the rim of the dish will arrive to the rear of the microphone. The size of the microphone can also have an affect on response; to reduce this it should have as small a diameter as possible.

Mounting the Microphone

Various methods are used for mounting the microphone in the reflector one being to use a clip which can be tightened around its body to hold it firm, the clip being carried on a rod directly connected to the dish. An alternative mount holds the microphone in suspension between stout elastic bands and has the two advantages of providing a barrier to the transmission of handling noise, and the ease with which the microphone can be put in position. It can accept almost any shape and size, and a well designed quick release ferrule allows for easy forward and backward movement to put the diaphram of the instrument on the focal point. It is doubtful if any mount would be one-hundred per cent effective against handling noise when high amplifier gains are in use.

Alignment of the Reflector

A reflector is very directional to high frequencies, a considerable loss occurring at 5 degrees off axis, and so it must be very carefully lined up on the sound source if full use is to be made of the gain provided. One method which is employed is to use a sight-hole, just off centre of the dish, through which a sight can be taken along the top of the microphone in order to line it up with the sound source. It can be reasonably satisfactory *if* the subject can be seen but this is frequently not the case – a bird can very easily be hidden by the leaves of a tree and stridulating insects lost in undergrowth. When the subject can be heard but not seen the sighting method fails but the sound can be used instead.

It is not difficult to see that when the strongest possible signal is being produced, the reflector must be correctly aligned. Signal strength will be indicated by the record level meter on the tape-recorder but many of these have dials on which small differences, which can be important, are difficult to see. An alternative method, which with practice can be very successful, is to monitor the sound being picked up by the microphone. By moving the reflector from side to side and up and down it is not difficult to determine correct alignment, in fact hidden birds and insects can be found by this process. The method requires a socket on the recorder into which a headphone can be plugged, it is preferable that the signal is available without the tape running and even better if the motor can be at rest – the latter saves a very considerable drain on the batteries. Providing suitable switches, or a pause control, are available it is then possible to start the tape running as soon as satisfactory alignment has been achieved.

Either a small, single earpiece or a complete set of headphones can be used for monitoring. The use of a single earpiece generally destroys the directional properties of the operator's ears, but if location is not found quickly through the equipment it is a simple matter to temporarily remove the earpiece, obtain the general direction of the subject with the ears, and then replace the earpiece for fine adjustment. The signal to the monitoring ear can generally be made loud enough to overcome the direct sound reaching the other ear. An earpiece has the advantage of being small and considerably less cumbersome than a set of headphones, whilst the latter do not destroy the ear's directional property and, providing no signal is being fed to them, some direct sound can still be heard, though this depends upon the insulating efficiency of the ear pads.

Sound reproduction in the headphones should be better than in an earpiece but this will be of no significance unless off tape monitoring is available; if monitoring is direct, then it can tell nothing of the quality of the recording being made. When off tape, A/B, monitoring is available then, undoubtedly, a set of headphones should be used whenever possible because they can provide a check on recording quality and a good comparison of the signal before and after it has been recorded, as well as being used as a check on direction.

2. Photo: Jack Skeel.
The author: Reasonably hidden and comfortable.

3. Photo: 3M
Richard Ballard using a Revox A77 studio tape-recorder to record the stridulations of an insect held in temporary captivity.

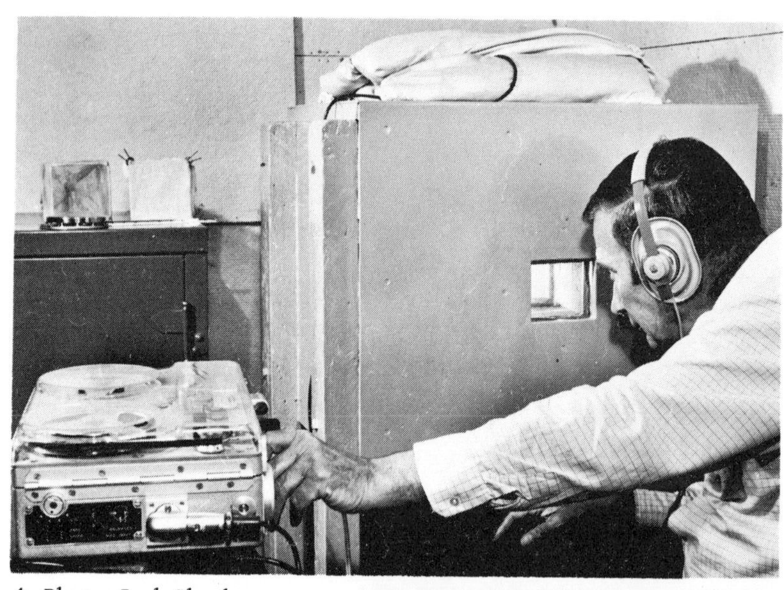

4. Photo: Jack Skeel.
Jack Skeel at work recording insect sounds with a Nagra 111 tape-recorder. To eliminate extraneous noise the insects are in a sound-proof 'mini-studio' with observation port.

5. Photo 3M
Jeffrey Davies using a home-made hydrophone and cassette-recorder to record underwater sounds.

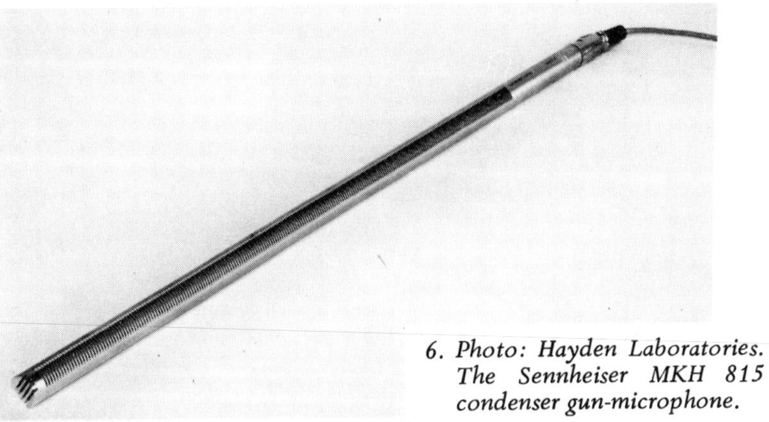

6. *Photo: Hayden Laboratories. The Sennheiser MKH 815 condenser gun-microphone.*

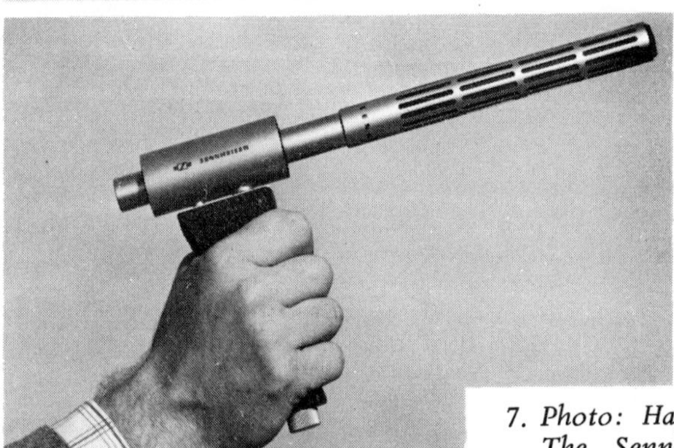

7. *Photo: Hayden Laboratories The Sennheiser MKE 802 electret short gun microphone*

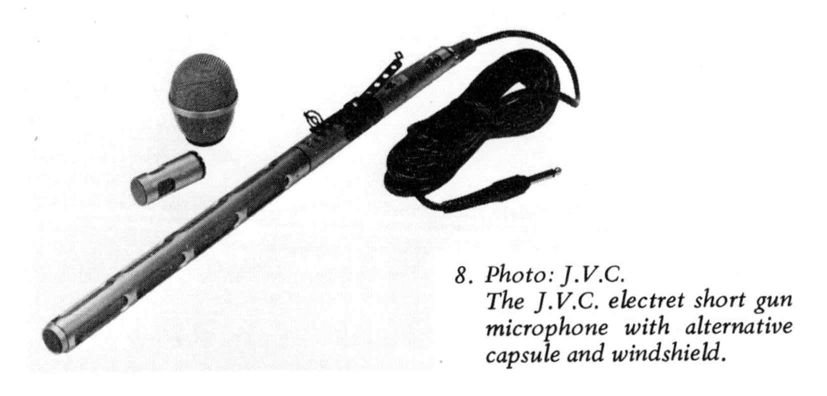

8. *Photo: J.V.C. The J.V.C. electret short gun microphone with alternative capsule and windshield.*

9. *Photo: Jack Skeel*
A chemical-retort clamp used to secure a microphone to a tree branch; the same clamp can be used to fix a microphone at low level by driving the spike into the ground.

10. *Photo: Jack Skeel*
The weight of the cable is taken off the soldered joints of the connector by the knot.

11. *Photo: 3M*
Comfort for long waits during day and night. Graham Burrough, with Edith Scourey, using Nagra I tape-recorder. The special platform was used during an extended study of badgers. Note the unusual, but very effective, way of 'mounting' the gun microphone.

12. Photo: 3M
Bill Sinclair with Nagra 1V-S tape-recorder and hand-held Sennheiser gun microphone.

13. Photo: 3M
Richard Savage using an A.K.G. type D190C directional microphone in a tripod mounted Grampian reflector of 24 inches diameter and 7 inch focal length, to a Tandberg Series 11 tape-recorder with cover removed to accommodate 7 inch tape spools.

14. Photo: Bill Bowles
Alan Ferry using a Fi-Cord 202A tape-recorder and a Grampian 24 inch reflector mounted on a monopod.

15. Photo: Bill Bowles
Jean Clamp using a light weight glass-fibre reflector of 20 inches diameter and 5 inches focal length, constructed to the author's design. The microphone mount, designed to reduce handling noise and to accept almost any microphone, also acts as a support for a 'bonnet' type wind shield. The tape-recorder is a Uher 4000 Report L with Grampian DP6 microphone.

16. *Photo: 3M*
    Douglas Ireland using a Sony TC 800 B tape-recorder and 20 inch glass-fibre reflector fitted with a 'bonnet' type wind shield.

17. *Photo: Richard Margoschis*
    Vivien, the author's wife, using an ITT : SL55 cassette recorder with 20 inch glass-fibre reflector and an ear-phone for monitoring. Note the right arm and elbow held closely into the side for support.

18. Photo: Sony
The Sony TC 153 SD battery operated stereo cassette recorder.

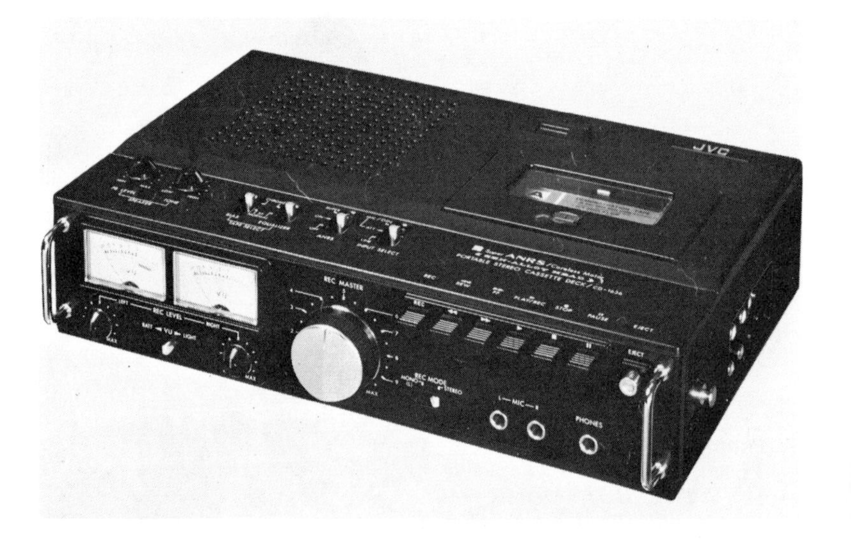

19. Photo: J.V.C.
The J.V.C. CD1636 battery operated stereo cassette recorder.

20. *Photo: Uher*
The Uher CR210 battery operated stereo cassette recorder.

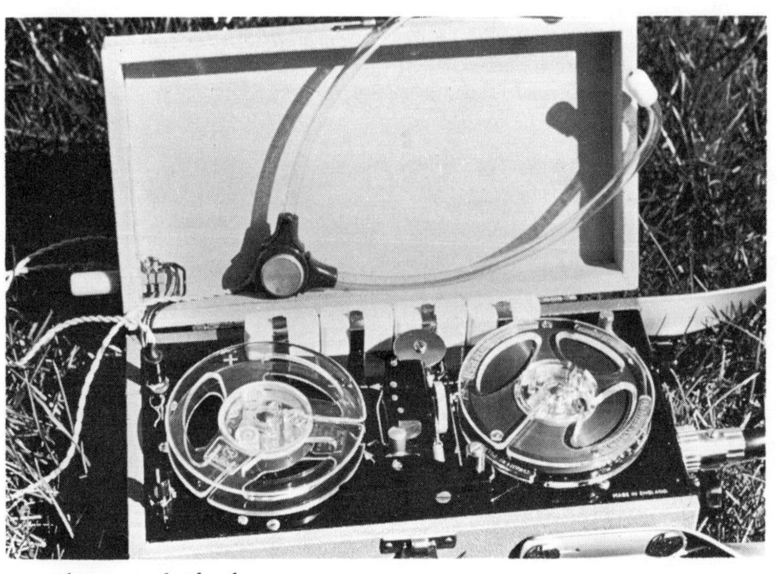

21. *Photo: Jack Skeel*
The Fi-Cord 1A battery operated tape-recorder. Now long discontinued, it was one of the first portable machines and weighed only 4 pounds.

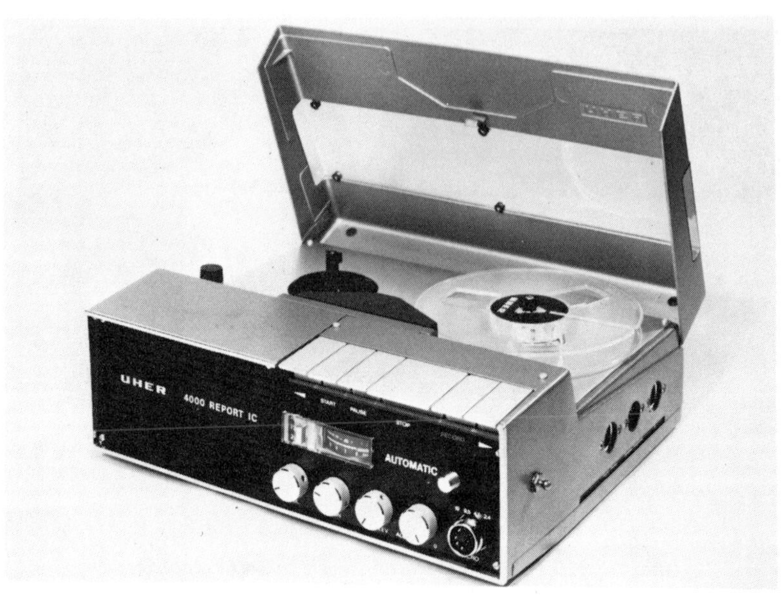

22. *Photo: Uher*
The Uher 4000 Report IC battery operated tape-recorder, available in mono and stereo versions; it has manual and automatic control, with tape speeds up to 7.5 inches per second.

23. *Photo: Jack Skeel*
The Tandberg Series 11 battery operated tape-recorder, has off-tape monitoring with tape speeds up to 7.5 inches per second.

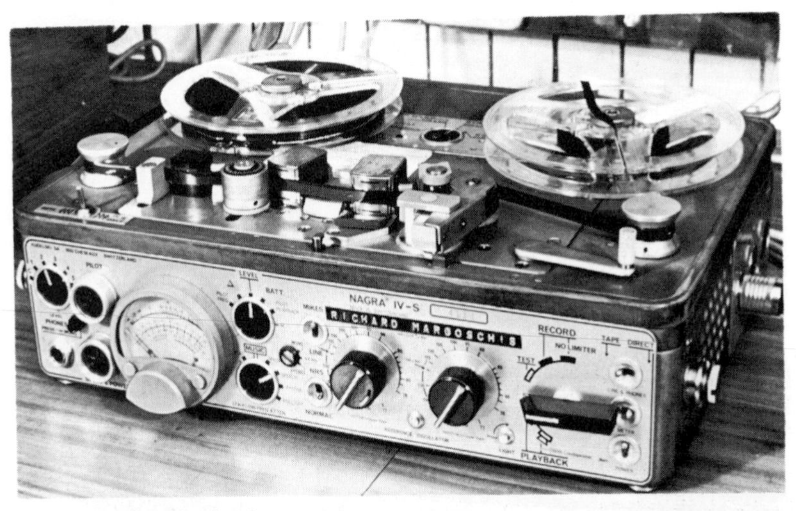

24. Photo: Jack Skeel
The Nagra 1V-S battery operated tape-recorder can operate in mono or stereo, with microphone mixing in mono, off-tape monitoring and tape speeds up to 15 inches per second.

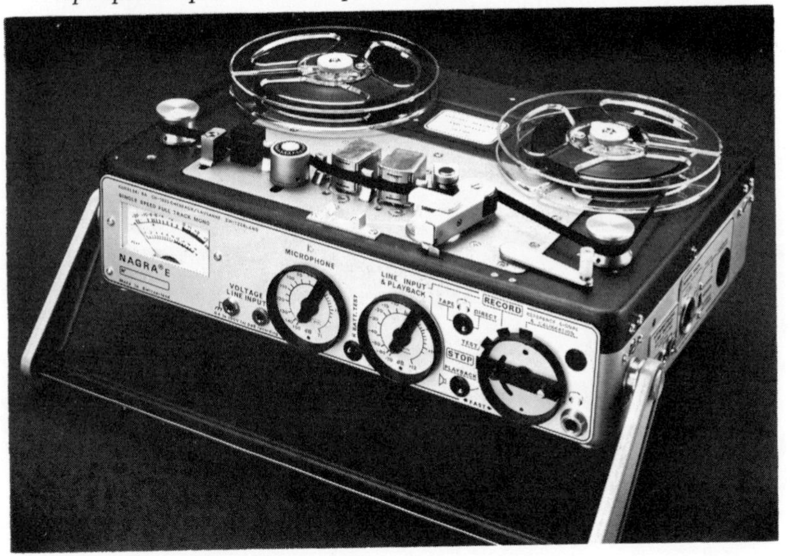

25. Photo: Hayden Laboratories
The Nagra E battery operated tape-recorder has off-tape monitoring with a single tape speed of 7.5 inches per second, full track.

26. *Photo: Jack Skeel*
A corner of the author's studio.

27. *Photo: Jack Skeel*
A studio stereo tape-recorder by Brenell with tape speeds up to 15 inches per second and capacity for 10 inch N.A.B. spools.

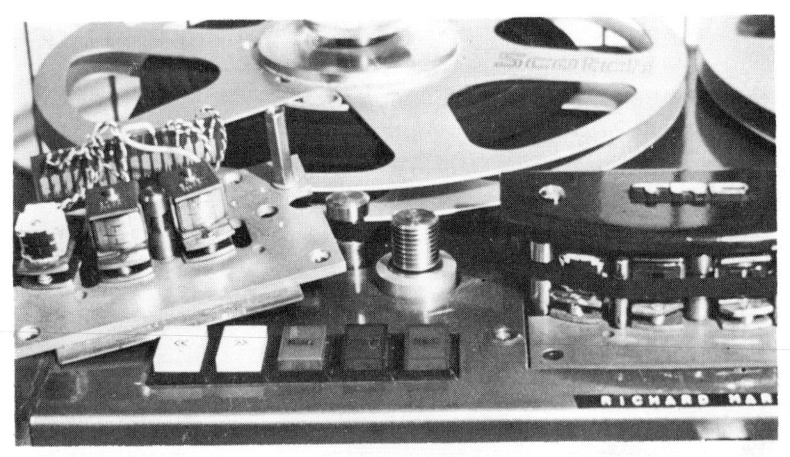

28. *Photo: Jack Skeel*
A section of a T.R.D. tape-deck fitted with a 'scrape roller' (arrowed). Placed on the deck, on the left, is an alternative 'head-block' which easily converts the machine from full-track mono to half-track stereo.

29. *Photo: Jack Skeel*
A mixer by Soundex, with four channels (two stereo pairs) having tone control available on two channels, and high and low pass filters on the other two; it can accept line or microphone inputs on each channel and provide phantom power for condenser microphones.

30. Photo: 3M
Bill Jackson building up a programme by track to track transfer on a pair of Ferrograph Series 7 studio stereo tape-recorders.

31. Photo: Bill Bowles
The author selecting a recording from the stereo section of his library of natural history sounds.

32. Photo: B.L.O.W.S.
Ron Kettle, sound librarian, at the British Library of Wildlife Sounds.

33. Photo: Bill Bowles
The author at work in the studio with a Revox A700 studio stereo tape-deck.

34. *Photo: Jack Skeel*
The author at work with a Beyer microphone mounted in his own glass-fibre reflector and a Tandberg Series 11 tape recorder.

Windshields

It is very rare that atmospheric conditions are such that the outdoor recordist does not have to take precautions against wind interference. Difficulties are produced in two ways; a strong breeze blowing through a wood, or even a hedge, can produce a considerable roar, and there is not a lot that can be done about it. Unless it is possible to select a position which will place the main source to the rear of directional microphones it is probably better to allow the sound to be a feature of the recording – perfectly natural in the case of subjects normally found on open moors and other windy places – but it must sound right. There is the argument that windy conditions are not right for making natural history sound recordings anyway, but against this is the fact that suitable protection against wind can result in satisfactory recordings being made in adverse conditions and can be of paramount importance when time at a particular location is limited, or when some very unusual occurrence is witnessed. Actually, the sound of wind can be very provocative and, if properly handled, will add another dimension to the recording; this is particularly so in the case of habitat recordings, and can have its place in connection with individual species recordings.

The other problem is that of wind 'blasting' produced by the wind blowing across the face of the microphone but, fortunately, there are ways and means of reducing its effect. Reference has already been made to the fact that a ribbon microphone can easily be damaged by the wind; the moving coil instrument is not so affected but, nevertheless, it does nor require a very strong wind blowing across its face to cause blasting. When a close microphone technique can be used – interviews, for instance – the record gain can be kept down sufficiently to eliminate a lot of the trouble, but in natural history work it is usual to have to use a relatively high gain and so the problem is increased.

Any form of windshield must be so designed that it will prevent the wind from blowing on the face of the microphone but not interfere in any way with the passage of sound. If a windshield does result in any attenuation, it will be the high frequency elements that are most affected; it must be acoustically transparent to prevent this. Most manufacturers offer windshields which are specifically designed to suit their microphones, and in some cases adaptors are available to allow one windshield to accept instruments of different diameters; their efficiency varies but few will allow the use of full gain in anything more than a breeze.

Generally speaking, the larger windshields are the most effective but, once again, the problem of portability arises. Ideally, an outer cover, as large as possible, is used to reduce the initial force of the wind, leaving a second shield, fitted over the microphone, to take care of any sudden increase in pressure occuring within the outer shield. A few such shields, specially designed to specific microphones, are available from specialist manufacturers but tend to be expensive; to the 'do it yourself' man they are not difficult to make.

The outer cylinder, at least 3½ inches in diameter, is made of ½ inch wire mesh; 'Twilweld' is very suitable and has the advantage that joints, made by wrapping cut ends around uncut squares, can be reinforced by means of soldering although this is not essential. This frame is covered, on the outer side, with a suitable acoustically transparent material fixed with impact adhesive. Caps, to cover each end, are made with a ring of Twilweld, covered with material, to a push-on fit and, for safety, held in place by means of 'Velcro' type fastener. A refinement is to apply a further covering of lightweight plastic ¼ inch mesh – as used in gardens – which can also be fixed with contact adhesive.

The microphone is carried on a light alloy strip to which is fitted a pair of 'Terry' spring clips, the latter being adjusted by the use of suitable 'stand-offs' to place the microphone

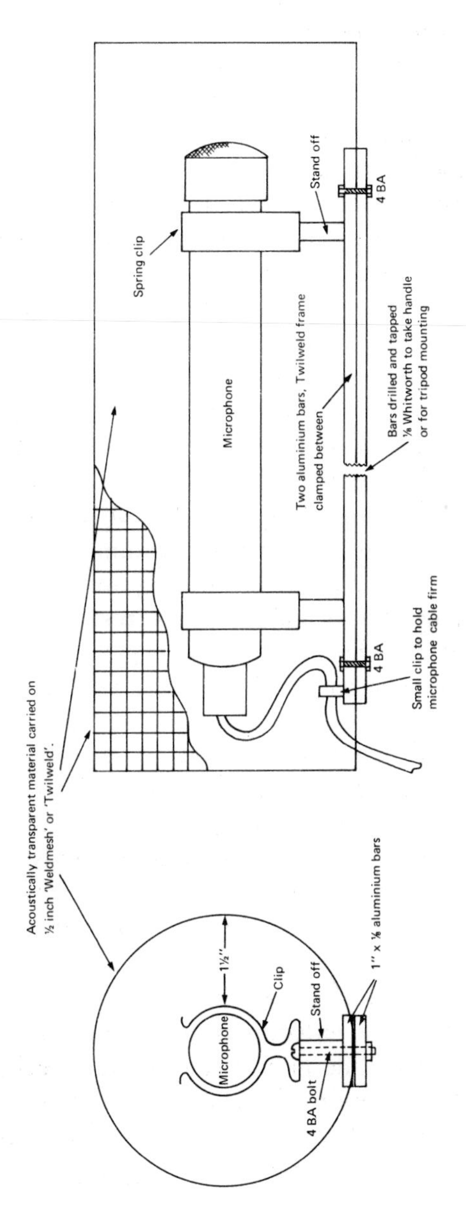

Fig. 8. Construction of windshield.

centrally in the cylinder. A second alloy strip, of similar dimensions, is placed outside the cylinder and clamped to the inner strip by means of two 4BA nuts and bolts, this outer strip is drilled and tapped with 1/8 Whitworth thread, at the point of balance, to allow fitting to a handle, monopod or tripod as described in connection with parabolic reflectors. An alternative method of mounting the microphone is to replace the spring clips with crossed pairs of strong elastic bands; the microphone is passed through and rests in the bands at the points where they cross. The method is useful in reducing handling noise but the microphone is not held so firmly and stability, within the shield, depends upon its weight and length.

If a stereo pair of microphones is to be used, then the shield will need to be considerably larger to accommodate them. The necessary spring clips to hold the microphones can be fixed to a base plate of either aluminium or plywood, and the wire mesh, to support the covering material, fixed to the base by passing the ½ inch length wires, which can be left at ½ inch intervals when cutting to length, through small holes drilled at ½ inch intervals. The shape need not necessarily be round.

The covering material must be acoustically transparent to prevent any loss. A special fully reticulated cellular foam, obtainable from specialist manufacturers and used to give an upholstered appearance to the front of loudspeakers, is ideal for the purpose. When using the cylinder type of windshield as described above, the second line of defence can be provided by making a sort of stocking, of the reticulated foam, which can be pulled over the microphone.

When a parabolic reflector is in use the rim of the dish and end of the microphone mount can provide ideal support points for a very large windshield made rather like a bell tent and using reticulated cellular foam; elastic is threaded through the circumference and provides sufficient grip to keep it in place.

The Choice of Tape-Recorder

Previous chapters should have given the reader an idea not only of what is expected of him, if he is to be the operator, but also what he is to expect of the machine he chooses to use. Basically, what is required is a machine which is capable of being taken into the field to record sound on tape; it must be self-contained to the extent of providing its own power supply from internal batteries. A very wide range of models satisfy these demands but the price range is so great that the first consideration must be a very personal one – what is the top limit of money available? As with most commodities, the facilities provided by the article, and its efficiency, are related to its cost. Many more points can now be considered to determine which machine is likely best to suit the requirements.

The nature of the work is such that the tape-recorder will need to be transported over considerable distances and, unless the recordist is satisfied by going only as far as a car will take him, it will have to be carried. Weight is therefore very important and sufficiently personal to depend upon the physical capabilities of the individual. The weight concerned is that of the machine ready for operation in the field, and must include batteries, carrying case and spool of tape; specifications must be carefully watched and checked on this point. The handle with which most machines are provided is perfectly satisfactory for carrying it around the home, or even short distances, but it is quite useless for fieldwork. It is essential to have a good strong carrying case which can give some protection against the weather and which includes a well fixed and strong shoulder strap in order to leave both hands free for other purposes. When in the case the machine should hang comfortably at the operator's side in such a way that it can be used in that position if desired.

The layout of the controls will often dictate whether the

machine is best used at the left or right side when hanging from the shoulder, but there are other very important considerations. All the controls should be arranged on the uppermost panel as the recorder hangs in its case, they should all be clearly visible and laid out in such a way that they are easily accessible. The three most important items are the record gain control, the record level meter and the pause control; difficulties can arise if the use of one interferes with another, for example, the fingers used to operate the gain control might intrude upon a clear vision of the meter dial. If the layout is satisfactory for operation from the shoulder, then it is fairly certain to be so if the machine is resting on the ground. The record level meter is very important, the larger its dial the better and it is an advantage if it can be illuminated at night; illumination should not be continuous as this wastes battery power.

A pause control allows the machine to be switched into the record mode and the signal from the microphone monitored with the tape stationary, as soon as recording is to commence the pause control is released to start the tape running. With some recorders it is possible to have the amplifier in the record condition, for monitoring purposes, without the motor running and thus effect a considerable saving on batteries because the motor alone generally accounts for about one-half of power consumption. Some recorders are provided with a position on the main control switch to allow for this, others, not so provided, can be adapted by the use of an extra switch plugged into the accessory socket. Control in this way takes fractionally longer to start the tape than is the case with the pause control.

The use of pause controls is often a means of eliminating noise from main operating switches, many of which make an unbelievable amount of noise when operated in really quiet circumstances. Mechanical noise can also be produced by the motor, many otherwise excellent recorders suffer to some degree from this fault; it becomes troublesome when conditions are ideally quiet and a hand held microphone, such as a gun microphone, or reflector set-up is being used.

Another source of unwanted noise is the electronics of the machine, it does not get recorded by way of the microphone, as does mechanical noise, but directly through the amplifier and, in some instances, it can include electronic interference from the motor. The performance of a tape recorder, in this respect, will be given in the specification as the signal to noise ratio and should be quoted for each speed at which the machine can operate.

The specification should also give the frequency response for each tape speed but it is of little value unless it gives the tolerances involved, these are normally $\pm$ 2 or 3 dB.

A point from which the signal can be monitored is an essential feature, and it should be situated in a convenient place. The type of monitoring which can be achieved depends upon certain features of the amplifier contained in the tape-recorder; the difference between the two-head and three-head systems is described in the section dealing with the techniques of making a tape recording. It is only the more expensive battery operated machines which employ the three-head system and, so far, no known battery operated cassette machine uses it.

Economy does not go hand in hand with quality, but quality goes hand in hand with tape speeds and track widths – as each is reduced the quality suffers. It is by no means an easy matter to record and reproduce the sounds involved in nature; as the quality achieved will be affected by these factors the aim should be to obtain the best within the limit of the money available.

The power supply is normally obtained from standard size dry cell batteries contained in a special compartment within the machine, the number of cells required to make up the power pack varying according to the electrical demands. An alternative is the re-chargeable cell specially made to fit into the battery compartment; its weight is about the same as that of the cells it replaces; A special unit is required to recharge these cells and it is sometimes built into the recorder resulting in some extra weight. A much better arrangement is for the charger to be separate but capable of being accommodated within the battery compartment when it is desired to run the machine from the mains supply. Another possible source of power is the car battery and provision to use this can be very useful.

The size of the tape spool which the recorder can accept will have a direct bearing on the maximum length of a non-stop run. The most popular size appears to be five inches which, when loaded with standard-play tape will give approximately a quarter-hour per track at 7½ inches per second, or about twenty-two minutes if loaded with long-play tape, the playing times being doubled at 3¾ inches per second and halved at 15 i.p.s. A number of machines designed for operating with five inch spools are capable of handling seven inch spools when the cover is open, this considerably increases the potential of the machine for use in the studio, especially when recordings are filed and stored away on seven inch spools. At least one battery operated machine is now capable of operating with seven inch spools when the cover is closed.

The first process in making a tape recording is to fit a loaded spool and thread the tape through the tape track, a process which must be repeated frequently under field conditions, maybe in the dark, and so the ease with which it can be accomplished is of some importance, and worth consideration when selecting a tape-recorder.

A means of determining how much tape is left on the reservoir spool is a great asset when the machine is being used in its carrying case; a suitable window is generally sufficient. If a digital tape indicator is provided it can be set at zero at the start of the tape then, providing the reading to be expected at the end of the tape is known, a good idea of the amount remaining is visible at any time.

Some of the more expensive tape-recorders include a 'limiter' which can be switched in or out of circuit. Its purpose is to handle any sudden loud noises which might, if not reduced by the circuit, cause overmodulation distortion. They can be useful in dealing with a bird which sings from a point rather closer than expected, but much depends upon the frequency range over which they are operative.

Other facilities provided are not essential to the process of *making* a recording in the field. For instance, any speaker built into the machine is a compromise and cannot produce the quality of a properly constructed speaker unit, but even a very small one is an asset in providing on the spot checks, although such checks can as easily be made by using headphones. Fast winding should be used as little as possible because it is a very heavy drain on the batteries.

Line in and out sockets, though rarely used in the field, add considerably to the versatility of the machine in allowing its use for recording a signal from another machine and, more important, passing a signal from it to another tape-recorder for copying purposes – the best machine to replay a recording is normally the one that made it.

Whilst these latter items are not essential so far as recording in the field is concerned, and tend to add both weight and bulk, they are points to consider if the field tape-recorder is to double for use in the home studio. No known tape-recorder is made specifically for recording the sounds of nature and so the final choice has to be, to some extent, a compromise. It might lack some desired facilities; almost certainly it will have some facilities that are not essential. A person who is handy at electronics and with instruments might well be able to carry out modifications to reduce weight and power requirements, such as the removal of a speaker, or to provide extra facilities such as a switch to stop the motor running when not required but retaining power to the amplifier.

Automatic gain control (AGC) and noise reducing systems are features which appear mainly on cassette recorders: they must be used with caution when making natural history sound recordings. Automatic gain control removes from the recordist the ability to dictate the level at which the signal is modulated when recorded on the tape. This sounds very helpful, as in fact it is for some purposes, but the evidence is, that it is not always capable of satisfactorily handling birdsong. As described elsewhere, the strength of the signal being passed to the record head must be carefully controlled if distortion due to overmodulation is to be prevented. When a tape-recorder is started with automatic gain in use the record level will be set, automatically, to deal with the loudest sound reaching the microphone but the system requires time, even if only very brief, to adjust to any changes.

Consider a location having the sound of a river in the background and where it is proposed to record the song of an individual bird fairly near at hand. While the bird is not singing the gain will be automatically brought up to handle the river noise, but when the bird suddenly starts singing overmodulation will occur until the a.g.c. has reduced the gain which, in turn, will affect the background level; the latter will remain constant while the bird sings, but if the bird is silent for a while the gain will 'recover' and be set again to the noise of the river. In the recording the noise of the river will disappear as the bird sings and reappear when the bird is silent – a very unpleasant effect. If the bird is very close to the river and his song is not strong enough to require a reduction in recording gain, then the result can be better but, nevertheless, there is difficulty in obtaining a good balance between song and background sounds.

A number of cassette tape-recorders are designed to rely entirely upon automatic gain control. From what has been said above it will be evident that such a machine is very likely to cause disappointment; one which provides manual over-ride to the automatic gain control is much to be preferred.

The various noise reduction systems found in many of the more expensive cassette recorders are designed to reduce hiss. Generally, if the system is used, it must be in circuit both for recording and replay, otherwise the result is unsatisfactory; it works automatically and, like a.g.c., can not always handle the transients in birdsong, with the result that changes in hiss level might be heard if the background is quiet. Provision is normally made to switch it out of circuit.

Open Spool v Cassette.

What has been said about the points to consider regarding the choice of a field Tape-recorder can be applied to each of the two basic types of machine – open spool and cassette. The pros and cons of each must now be considered.

Without doubt, the first half of the 1970s has seen a vast improvement in the quality of recordings obtainable on cassette as a result of developments affecting both the machine and the blank cassette tape. One of the results of this advancement has been a serious reduction in the choice of open spool battery tape-recorders, indeed it has had a similar affect on lower priced mains machines, whilst the choice of cassette recorders has widened considerably.

The standard compact cassette tape-recorder employs the very low tape speed of 1 7/8 inches per second, and the tape, on which two mono or four stereo tracks are laid, is approximately 1/8 inch wide instead of the ¼ inch used by open spool machines. Because of these limitations the cassette still can not match the quality of a good open spool machine, particularly where natural history sounds are concerned.

At one time it was fairly safe to say that the cassette machine had a distinct advantage over the open spool in that it was smaller and considerably lighter; whilst this is still true of some models, many have tended to become as large and heavy as some open spool machines. However, cassettes have retained their cost advantage, both in respect of machine and tape. On the other hand, because it is so small, engineering problems arise and it is doubtful if the life of a cassette machine is as long as that of the open spool type.

It is not practicable to edit cassette tape. If any editing is to be carried out – and a keen recordist will certainly want to do this – then the recording must be copied on to open spool for the work to be done in that form. Similarly, for storing away the sections of recordings to be kept, they must be copied on to either open spool or another cassette so that they can be arranged in a reasonable order. If this is not done, an awful lot of unwanted material will go into storage and re-useable tape lost. There is a very considerable saving on required storage space; two hours of programme material on standard play open spool tape recorded ½ track at 3¾ i.p.s. requires approximately 36 cubic inches, against approximately 17 cubic inches for the same programme on two C60 cassettes.

The compact cassette system has an advantage in the ease with which loading is effected; there is no tape threading to be done because it is necessary only to slip the cassette into a compartment in the machine.

New cassette systems using ¼ inch tape running at 3¾ inches per second are now beginning to appear. They will, of course, produce a new generation of tape-recorders, and it remains to be seen whether or not they equal the success of the compact cassette.

Irrespective of what has been said, the cassette system has resulted in a lot more recording being done outdoors and very good results are being obtained; if limitations in choice dictate 'cassette or nothing' then, certainly, a cassette machine – of as high a quality as possible – should be purchased for the work.

Tape Recorder Maintenance

The day to day maintenance of a tape-recorder, cassette or open spool, does not require any special knowledge in electronics or mechanics but it is a very important part in the production of good recordings.

Probably the most important feature is cleanliness. As recording tape passes through the tape track it is inevitable that a certain amount of the coating rubs off, it sticks to the tape guides, it builds up on the rollers and it falls on the deck. Special cleaning fluids are available and should be applied by means of cotton wool on a stick. Attention should be given to all tape guides, rollers and capstan and also to the face of the heads, extra care being exercised when attending to the latter in order not to disturb them or damage the polished surface. Any dust lying on the surface of the deck can be removed by a soft brush.

Under no circumstances should oil be applied: the slightest amount of oil or grease is liable to cause tape slip and, consequently, irregular speed of the tape.

From time to time the tape heads require de-magnetising to remove any build-up of magnetism which might have occurred. This can be done quite easily with a special instrument which is connected to the mains supply; the method is to bring the tip of the instrument close to the head and then with the current still on, to draw it slowly away.

Dirty heads can cause the loss of high frequencies with the result that recordings do not seem as 'bright' as they should do; if cleaning does not provide the remedy then the chances are that the head may have gone out of alignment or be badly worn, and the machine should be checked over by a technician who has the necessary equipment.

Magnetic Recording Tape.

Magnetic recording tape is a very thin plastic base, one side of which carries a special magnetic coating, it varies in thickness according to type but its width, for open spool, is standardised at approximately ¼ inch which must be maintained within close tolerances, otherwise the tape will not run smoothly through the tape track of the tape-recorder – it is very important that it should do so.

The thickness of the tape dictates the length which can be carried on a spool of given diameter and, therefore, has a direct bearing on the duration of recording time available. The thickest tape is known as 'Standard-play' and 600 feet of it can be carried on a 5 inch diameter spool; used at a speed of 7½ inches per second it will have a duration of 16 minutes per track. The same spool will carry 900 feet of 'Long-play' or 1200 feet of 'Double-play' tape, with a corresponding increase in recording time. A 7 inch diameter spool carries twice as much tape as a five inch one.

Standard-play tape is obviously stronger and not so easily damaged as the thinner ones, long-play is perfectly satisfactory and, in some cases, might allow better contact between tape and record head, whilst 'double-play' tape should be resorted to only when an extended continuous running time is required. The same remarks apply to cassette tapes; the tape in a C60 cassette will have a running time of 30 minutes per track (total running time 60 minutes) and is twice as thick as a C120 having a running time of 60 minutes per track.

Some tapes, generally the thinner ones, are likely to suffer more from 'print-through'. A strong signal recorded on the tape can be transferred to the layer of tape lying above and below it, but at a much lower level, causing the effect of pre and post echo.

There are other differences; the most important is probably the type of coating used. In order that an undistorted signal can be recorded, the record amplifier provides a special bias signal and the optimum amount of bias required varies from tape to tape. Before a tape-recorder leaves the factory, the bias will have been accurately set up to a specific tape. Low noise tapes tend to require rather more bias than others and so it is as well to find out, when purchasing a recorder, the type of tape for which it is set up. If it is desired to use a different type, then it should not be beyond a good dealer to have the bias adjusted, if necessary.

The situation is rather more complicated in respect of cassette tapes because three main types of coating are used. The most common is 'ferric', so called because of the iron based material used in the coating; 'chrome' tapes have, as the name suggests, chrome as a base, and 'ferrichrome' have a mixture of each. Chrome tapes can accept rather more high frequency signal than ferric and provide a better signal to noise ratio, but for full advantage of this the correct bias must be used. Modern cassette recorders have a switch which should be adjusted to the type of tape in use; if no such switch is provided then there is little point in using chrome tapes but some of the ferrichromes might give an improvement over the ferric tape. A more recent introduction is a range of super-ferric cassette tapes which give a wider frequency range with the normal ferric bias and, possibly, less wear on the record head than chrome tapes. Some high quality cassette recorders also have an 'equalisation' switch; equalisation is tied up with bias and the two switches should be used as recommended by the manufacturer for best results with a particular type of tape.

Any imperfections in the coating of the tape can cause temporary loss of signal, either total or partial; such a loss is referred to as a 'drop-out'. The effect is of very short duration but quite distracting, especially in stereo recordings when only one channel may be affected. No matter how small the imperfection, it is aggrevated by slow tape speeds and narrow tape-tracks; doubling the tape speed reduces the duration of the drop out by half, and double the track width – ¼ to ½ track, for instance – can reduce its effect. Another point in favour of higher tapes speeds, wide tracks and good quality tape.

Tapes, both virgin and recorded, can safely be stored in normal home conditions where temperatures are reasonably constant, but they should not be left where they can be influenced by any stray magnetic fields, for instance, they should not be placed on, or too near, a speaker cabinet.

Processing

The original, or master, of any good tape-recording is an important, and probably valuable, item; it should be properly indexed and filed away in a suitable system for storage, as described in a later chapter. If the recording is to be sent to anyone, or used for programme or competition purposes of any kind, then it must be copied. There is inevitably some loss of quality whenever anything is copied but if good class equipment is used, and the job is done correctly, then the loss to a first generation copy should be virtually undetectable.

The process of making a copy is very similar to that of making a live recording, the microphone having been replaced by another tape-recorder. One machine is used to reproduce the signal and pass it on to the second machine by way of a suitable connecting lead between the line-out and line-in sockets of the respective recorders. A serious loss of quality will occur if a copy is made by placing the microphone of one machine in front of the speaker of the replay machine.

The two tape-recorders having been connected, a test run should be made of the tape to be copied in order to determine the correct gain to be applied on the copying machine. This done, the record gain is brought back to zero and the tape on the copying machine started, a moment later the master tape is started and the record gain control turned up to the predetermined level; as the recording comes to its end, the gain control is returned to zero. This procedure should produce a copy without any plops at beginning and end. Correct modulation of the copy tape is very important – the results of incorrect modulation are described elsewhere.

It is at this point that a further difficulty might make its presence quite evident, even though it can be present in the original. Especially if the subject contains dominant high frequencies, as is the case with many bird songs and other similar sounds, the copy just does not sound as 'clean' as the original. The trouble is a form of distortion which appears to result from modulation of the signal due to a high frequency vibration of the tape as it passes through the tape track. Some tape-recorders are more prone to the problem than others, and dirty pressure pads and tape guides are likely to add to the effect. A free running roller, over which the tape must pass, can be effective in considerably reducing the distortion produced. It is sometimes possible to add such a roller in the tape track if one does not already exist, it should be before the record-head and as near to it as possible. The term 'scrape roller' has been used to describe such a device – 'scrape' being used to describe the distortion.

Various modifications can be made to a recording during the copying process. It can, for instance, be reproduced at a different tape speed, but it must be realised that to copy at a higher tape speed will neither replace losses occurring during the original recording nor will it improve the existing signal to noise ratio; it will, however, reduce the likelihood of introducing further distortion and maintain a good signal to noise ratio in the copy – it is always worth using as high a tape speed as possible if quality is important, especially if further copying has to be carried out before the final result is achieved.

Processing can not make a bad recording into a good one but, in some instances, a poor one can be improved, sometimes to the extent of making it acceptable. Modifications are made electronically and mechanically; filtering is an electronic process, editing is mechanical and a change of speed is really a combination of the two.

Electronic filtering can be used to remove unwanted noise such as traffic roar. If the recording is satisfactory in other respects, then it could well be worth using a steep cut bass filter to remove the roar, a cut at around 500Hz should be adequate. This filtering process can be applied in the microphone line when the recording is made, in fact some of the more expensive tape-recorders include a variable cut filter in the microphone input circuit but, because great care has to be exercised in the amount of cut applied, it is better done by inserting a suitable filter in the line between the two tape-recorders when copying – test runs can then be made without spoiling the originals. Steep cut filters should be capable of switching to cut below different frequencies, usually 250Hz; 500Hz; 1KHz; 2KHz and 4KHz – the cut applied being 12 dB per octave with the roll off point at the frequency setting.

Tone controls do not have the same effect as steep cut filters; they are used to reduce or emphasise bass and treble. Such controls provided on a tape-recorder operate on replay only and, in any case, line-out is usually taken from a point in front of them. The output to an extension speaker, on which they do operate, is not a satisfactory point from which to take a signal for re-recording.

The 'graphic equaliser' is a considerably more complicated piece of equipment than either steep cut filters or tone controls, and might be looked upon as a combination of both. It can be used to either cut or emphasise frequencies throughout the audio spectrum and operates on pre-determined octave bands. A filter set at 2KHz will have an attenuating effect on all frequencies below that point, but a graphic equaliser could be used to either attenuate or boost frequencies in the 2KHz (or any other) octave band without seriously affecting frequencies above or below. In addition, the amount of attenuation or boost given is variable, usually within the range of ± 12dB. The principle use of the instrument is in connection with recording music and its application to natural history work may be limited, but its versatility as a tone control/filter can prove useful in connection with unwanted sounds.

The amount of cut which can be applied depends principally upon the contents of the original. If the recording to be processed is one of the stridulations of a grasshopper it is more likely than not that a cut from 4KHz down will have little affect on the frequencies present as a result of the stridulations but any rumble present on the original will disappear; on the other hand, if the song of a blackbird is involved then, after such a cut, it will not sound right for the simple reason that, as well as removing the rumble, some of the bird's voice will have gone also, and if the subject was a bittern – he might well have disappeared altogether!

A heavy bass cut of this nature is not so successful if the original recording contains the sound of wind in trees, or some similar noise having a wide frequency band, because the removal of the bass frequencies has the effect of accentuating the high frequencies to produce a 'hissy' result.

A habitat, or atmosphere recording will not sound right if it is filtered too heavily because it can very easily be made to sound 'thin' and lacking in tonal quality. However, if a recording of an individual bird is to be mixed into such a habitat recording, the bird recording can be filtered so that when the two go together there is no appreciable increase in background noise. The safest procedure when doing any processing of this nature is to do two or three tests using different degrees of cut.

Editing

The art of editing involves the physical removal of unwanted sounds and, possibly, the re-arrangement of sounds which are still to be part of the recording: the process also includes the arrangement of a series of recordings to build up a programme, and the removal of recordings for filing away as library items. The tools required are quite simple; a splicing block, a chinagraph pencil, sharp razor blade and a roll of special jointing tape.

Splicing blocks vary in design, some having clamps to hold the tape in place, but the most simple and easy to use is that which has a specially shaped channel into which the tape is inserted by running a finger along the tape as it rests over the channel – only light pressure is required. A chinagraph pencil is one which will produce a mark on a polished surface and is used to mark the exact spots at which the tape is to be cut with the razor blade; yellow or white shows up best on most tapes.

The jointing tape is specially manufactured for the purpose, any ordinary adhesive tape *must not be used* because it is liable to cause successive layers of tape to stick together when spooled up. A jointing tape 7/32 of an inch wide is just right to allow for the easy laying of the tape on to the recording tape as it lies in the groove of the splicing block.

When the two marks have been made on the tape, to indicate the points between which the edit is required, the first length of tape is pressed into the splicing block in such a way that the mark it carries is above the thin grove which will guide the blade as the cut is made. The second piece of tape is then applied to the block, above the previous one, again with its mark against the cutting groove; this brings the two extremities of the edit into correct relationship. The two tapes are then cleanly cut by a decisive sweep – rather than a press – of the blade; this results in four pieces of tape being held in the block, and one unwanted piece – the top one – is removed. The joint is now ready for completion by the application of a piece of jointing tape about one inch in length. Adequate pressure should be applied before the tape is removed from the block with a twisting motion, and the joint finally pressed hard to ensure adhesion of the jointing tape; a good joint should show no white between the ends of the cut tape, and no overlap of tape at the point of joining. The jointing tape is *always* applied to the back of the magnetic tape, not to the coated surface which runs against the heads. A little french chalk applied to the tips of the finger will prevent skin grease and moisture building up on the tape during the process.

The majority of the editing done to natural history sound recordings is a 'cleaning-up' process to remove any plops or the occasional unwanted noise of short duration, such as a motor car horn or a human shout; the amount of this work that a recordist will be prepared to carry out depends upon the value of the recording. Re-arrangements of the sounds in the recording must be treated with great care because it might be interfering with the natural sequence of calls of an animal and so ruining the authenticity of the record; it would be quite wrong, for instance, to reduce the duration of silence between the calls of an animal simply to reduce the overall duration of the recording; unfortunately it is, at times, done in the best of circles! If, for any reason, such modifications have been made then the fact should be stated.

An occasion when some re-arrangement is certainly permissible is when the recordist has been caught unawares – which happens only too often – and the tape is started only just in time to record some special sequence. The tape should always be allowed to run for a short time after the sequence has ended, it is then possible, if it becomes necessary, to cut out a few seconds of 'atmosphere' at the end and splice it in at the beginning to provide a decent 'run-in'. If the recording starts in the middle of a song sequence which is repeated then, of course, the incomplete sequence should be removed. Care must be taken to see that backgrounds match properly at editing points.

A considerable amount of practice is required to find the exact spot at which a cut is to be made, and this process becomes more difficult with the slower tape speeds – very accurate editing can be achieved at 7½ inches per second but it is twice as difficult at 3¾ i.p.s! Assuming that a plop is to be removed, the process is to stop the tape the moment the sound is heard; the point is then just a little past the replay head. The recorder should now be switched to a condition in which the amplifier is still alive and the tape can be moved gently to and fro by one hand on each spool; in this way the plop will be heard passing across the head, and if the degree of movement is gradually reduced the point can be placed exactly at the centre of the head and a mark applied on the back of the tape. A cut is then made each side of the mark and the tape rejoined – after removal of the bit carrying the plop. Experience will tell how long a piece to remove; it is always best to remove too little in the first place because, if some of the unwanted noise remains, the joint can easily be broken and more tape removed.

It will be necessary to carry out some experimentation to

discover the best way to do this with any particular machine. Difficulty can be experienced in getting the tip of the pencil to the replay head because of covers, if these cannot easily be removed then a more convenient spot should be selected and the exact distance between it and the head measured out along the edge of the splicing block, so that the pencil mark can be placed against it – instead of the cutting groove – when applying the tape to the block.

The beginning and ending of longer sequences, which are to be joined, are found in exactly the same way and the necessary marks made.

Mixing Recordings

The mixer is a device which allows two or more signals to be mixed together at varying levels, as desired, the composite signal being available for transmission either for replay or re-recording purposes. The number of signals that can be handled at any one time depends upon the number of 'channels' available because each signal occupies a channel.

In its most simple form the piece of equipment is no more than a series of resistors and a potentiometer to act as a gain control on each channel. The form has the disadvantages that a certain amount of signal power is lost and there is likely to be some interference between channels, but no power supply is required and the unit can be very small and compact. The disadvantages are overcome by providing an amplifier to each channel giving individual level control; the level of the mixed output signal is determined by the master control, which is included in the circuit of the mixing amplifier, to which each line amplifier feeds its signal. The mixer might be such that it is capable of accepting signals from microphones only, but most provide line inputs on some, if not all, the channels.

If constructed to specification it is worth incorporating high and low pass steep cut filters and tone controls, each available on certain channels, within the same unit. The provision of providing a power supply from batteries, as an alternative to a mains unit, is useful in that it allows the unit to be used in the field when it is desired to use more than one microphone.

Presentation

When a recording is to be sent away, whether it be to a friend, a prospective purchaser, or for competition purposes, it will make a much better impression if it is properly presented.

A white leader tape of adequate length – a minimum of thirty inches – should be spliced on to the beginning; this tape can carry the name of the author and other relevant details such as tape speed, subject matter and so on. The recording should start immediately after the leader and appear with a short but positive fade up; there must be no plops. The finish should be a similar fade off followed by a adequate length of red trailer tape. Spools carrying the recording should also bear an indication of the subject matter and source of origin, together with tape speed and approximate running time.

If it is likely that the person receiving the recording will want to copy it, for insertion in a programme for instance, then it should be presented at the highest tape speed possible. The recording should be as fully modulated as possible – consistent with distortion – and it can help if it is preceeded with a few seconds of 1KHz tone to zero modulation; this reference tone tells a recording engineer at what level a recording should be modulated when copying – if it is not there, he will find out for himself.

Building up a Programme.

Once a library of natural history sounds has been built up, a lot of pleasure can be derived from weaving them into programmes or building up composite recordings; a programme consisting of a series of recordings linked by a narrative can be produced with the minimum of equipment. Two tape-recorders are required in order that the library items can be copied but a field recorder can serve as replay machine in conjunction with a mains machine.

The easiest method is to record the narration on one tape, leaving a short space at points where recordings are to be inserted; the recordings to be used are then copied on to another tape. The process now is to mark in turn the beginning and ending of the spaces between the narration and of the recordings, the latter is then spliced in between the two pieces of narration. The recording should have a quick fade in and out and should be cut in close to the ends and beginnings of narrations, any unwanted plops being edited out in the process.

Mains operated tape-recorders often have the facility of mixing a microphone input with a line input, so providing the opportunity of a different approach. All the recordings are spooled up in the required sequence and of correct duration, which may mean copying first, and separated by a short length of coloured leader tape to provide a cue. The equipment is then set up with the replay machine feeding into the line input of the recording machine and a microphone into the microphone input; record levels for the microphone and *each* line recording are then checked and noted. With the replay machine ready to go upon the release of the pause control, the record machine is started and the microphone gain turned up to receive the first piece of narrative. A few moments before this is due to end, the first recording to be inserted is started by releasing the pause control and the line gain control brought up to the pre-determined point; as the coloured leader appears, to provide the cue, narration is re-commenced, the line input gain returned to zero and the tape held on pause until required to produce the next insert. This method requires an open microphone which, so long as its input gain is turned up, will pick up any sounds made in operating the equipment and considerable practice is required for one person to act as both recording engineer and narrator. An alternative is to have a colleague to present the narration to a microphone placed at a safe distance.

An effective use of a composite recording is to present, within a pre-determined time limit, all the calls of bird, mammal and insect species which have been heard in a given habitat during a certain time of the year. Because it will be necessary to mix several recordings together, much will depend upon the equipment available. With little more than one mono and one stereo tape-recorder, a programme can be built up step by step, providing the stereo machine is capable of track to track transfer. A stereo recorder can be described as two mono recorders having a common tape transport system, and many such machines provide the facility to transfer a recording from one track to the other track whilst mixing it with an incoming signal from another machine.

A suitable recording is selected to provide continuity of action and atmosphere throughout the programme to be produced – perhaps three or four minutes – and is first copied from the mono machine on to track two (bottom track) of the stereo machine. The recording to be mixed with it is then replayed from the mono machine to track one (top track) of the stereo machine, with the latter switched to transfer track two to track one – the two recordings will then be mixed together on track one of the stereo machine. If a further recording is to be added, the procedure is repeated by transfer of the new track one to track two, whilst adding the third recording from the mono machine. The process can be continued until all the required sounds have been added, but it must be appreciated that some loss of quality will occur as each transfer is made, and transfers should be so arranged that the final recording is made on track one (top track) of the stereo machine in order to be compatible with a half-track mono recorder.

If track to track transfer is not available, the build up might be accomplished by copying two recordings on to separate tracks on the stereo machine and then replaying them together to the mono recorder. There are many variations on the theme and much depends upon the facilities provided by the equip-

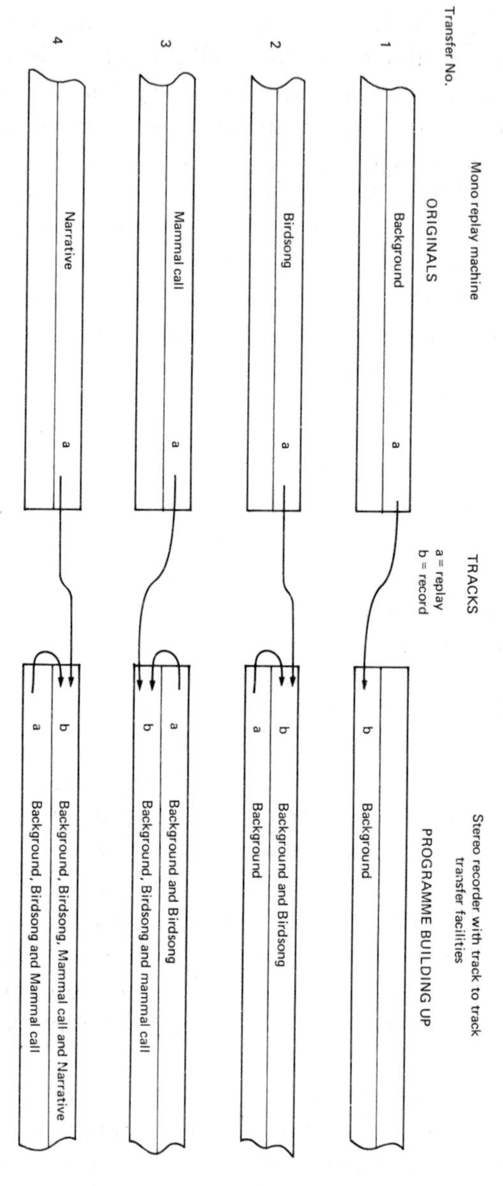

Fig. 9. Track to track transfer to build up a programme. ½ track recordings are illustrated.

ment available.

Care must be taken in selecting the recordings which are to be mixed into the basic background to see that their own background does not intrude, if it does, each individual recording will be noticeable as it comes in and the effect will be ruined. On this point, correct mixing levels are important; for instance, if one recording does have a noisy background it can possibly be used at a sufficiently low level, or filtered, to prevent the unwanted sounds from interfering – the calls concerned may appear to be distant but that can be perfectly natural in a habitat recording.

If all the recordings to be used have, in fact, been recorded within the habitat and season concerned, there should be little fear of introducing an error, but difficulty might arise if it is known that a certain species inhabits the area and no recording has been made of it. A suitable recording made elsewhere can be introduced but great care should be taken to check documentation to see that the call really does fit the requirement.

Replay of Tape Recordings.

In striving to obtain the highest possible quality from the equipment available the problems of reproduction of the recorded signal are often overlooked. The electronics of a tape-recorder should be capable of reproducing the full quality of a recording made on the machine, but can the speaker handle that quality? When a speaker unit is built into the same cabinet as a tape-recorder, it is, of necessity, a compromise, and the recorder should have an output socket to which an external speaker can be connected by insertion of a plug which will also mute the internal speaker. The improvement in quality achieved will, naturally, depend upon the capabilities of the external speaker unit and, as with most things, cost plays an important part, but care should be taken to see that the speaker properly matches the output. When a complete high quality sound reproduction system is available, the most convenient method is to take a lead from the line output of the tape-recorder direct to the tape input socket of the replay amplifier.

There is often a tendency to replay natural history sound recordings at too high a level. When heard in the field these sounds are at a relatively low level; to hear a bird singing in a room as if it is perched on the shoulder is not really natural. An idea of the level at which they should be played can be obtained by playing a recording of a well known piece of music – or other suitable recording – with the replay gain set for normal listening; without altering the settings a recording of birdsong, which is known to be well modulated, is played – the result can be quite revealing.

Natural History Sounds in Stereo

It must be remembered that, whilst stereophonic sound can give better realism and more information than mono sound, it is still an illusion. The sound, instead of being reproduced by one speaker, is emanating from two speakers and produces a 'sound picture' across the plane between these two points – it must not appear to come from two separate points; there will be an optimum listening zone outside which the stereo picture will deteriorate. The sounds which produce the illusion are from two separate, but related, recordings which mingle in the space between the speakers and the listener's ears, the recordings having been made at precisely the same time but kept in quite separate 'channels' from the microphones to the speakers.

When heard on headphones, the effect is rather different because the 'mixing' takes place within the listener's head instead of in space.

What can stereo do for natural history sounds? It is probably most effective in respect of habitat or atmosphere recordings; instead of the sounds from all the species present coming from a single point, and so being superimposed one upon another, they are spread out across the sound picture and can be identified as individuals, even when they are calling at the same time. Therefore the requirement, for a good mono recording of a habitat, that the calls of individual species should be spaced out so as to be clearly identified, is not so important when the recording is made in stereo. Movement is not necessary, but when present can be very effective, especially if it is kept within the sound picture.

In stereo recordings of individual species the background becomes very important. In the case of an individual bird, the balance between the background and his song must be such that the required song stands out clearly, the other songs will then

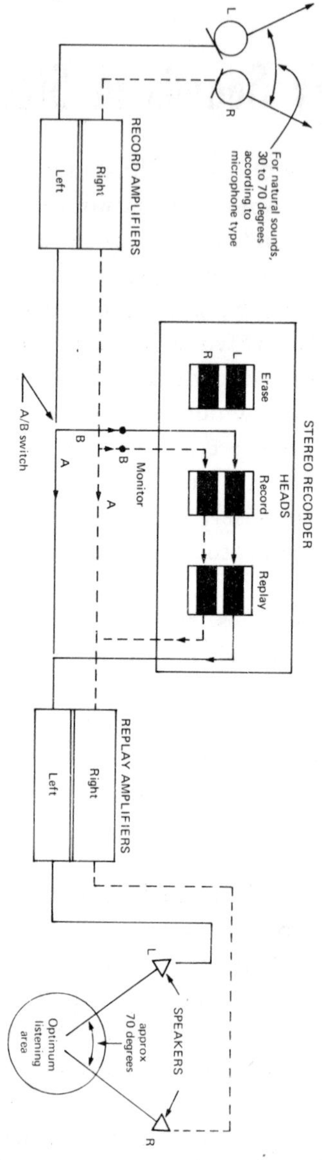

Fig. 10. Diagram of Stereo System; ½ track, 3 head system providing A/B monitoring. In a two head system an amplifier in each channel would perform both record and replay functions, and A/B monitoring would NOT be available.

be heard to each side of him, instead of all behind him as would be the case in mono, and any movement he makes will be related to the background sounds. Should the microphones be too far away from the bird, then his song will be at a similar level to the others and he will be lost amid them; on the other hand, if too close, the background will disappear and a clinical recording will result – with no background, and no movement, the recording may as well be in mono. The situation might be quite the reverse if something like a fox is being recorded on a very quiet night; 'clinical' background is to be expected and, whilst even a slight sound of wind in trees would place the animal in a stereo setting, his movement across the picture, while barking, would be ideal.

If a bird is being recorded when singing from a song-post there may be an opportunity to decide where to place him within the stereo sound picture, by directing the microphones accordingly. It is probably most natural to set him in the centre, and this is a very good idea if he is likely to move about, for it gives leeway either side; but there is no reason why he should not be set to one side providing the background is sufficient to 'hold up' the other side.

Special equipment is required to record and reproduce in stereo, but this same equipment can be used both to record and reproduce mono. Remarks which have been made elsewhere in this book regarding the selection of suitable equipment apply, in general, to stereo, and the choice available is not very much reduced because there are stereo versions of most open spool field recorders now available on the market, and the range of battery operated stereo cassette machines is improving. The cost is somewhat increased but weight and overall size are hardly affected; there are, of course, extra controls to look after! Basically, then, the only additional piece of equipment to be taken into the field is another microphone.

Directional microphones are very suitable for use as open microphones over a limited range; they should be mounted on a carrier and care taken in respect of the angle between the axes. If the angle between them is too great the result will be 'bunching' of sounds at each loudspeaker, leaving a 'hole-in-the-middle'; an angle of between 40 and 70 degrees between axes is a starting point for tests for any particular pair of such microphones.

Gun microphones, particularly the radio-frequency condenser type, are ideal for medium range work; mounted side by side, and in efficient windshields, the angle between axes should be kept down to about 30 degrees.

Microphones mounted as a coincident pair – with their diaphrams one above the other – can reduce phasing problems, particularly in the higher frequencies, which sometimes result in the subject's voice appearing to come from more than one point in the stereo picture.

The parabolic reflector, so successfully used by so many recordists for so many years, is not to be discarded in connection with stereo recordings; it can be adapted. Of the various methods tried, the most successful appears to be the division of the reflector vertically down its centre by means of a baffle, for which either plywood or glass fibre can be used. A microphone is fixed, at the focal point, on either side of the baffle and each having the benefit of half the reflector. Because of the reduced reflecting area for each microphone, it is an advantage to use a reflector of rather larger diameter than has been suggested as suitable for mono recording, nevertheless, a twenty inch divided reflector has been used successfully, even if over a rather reduced range. As with mono recording, the reflector's angle of acceptance is tighter than even a gun microphone, thus making it suitable for longer range recording but the feature can also be useful for making an individual species more prominent in a stereo background.

When used at close range, phasing problems can occur in the high frequencies and, upon replay, have the effect of making it sound as if the bird is singing from two different positions. In addition, the reflector's loss of low frequency sound still applies and can cause difficulty in 'placing' such calls within the sound

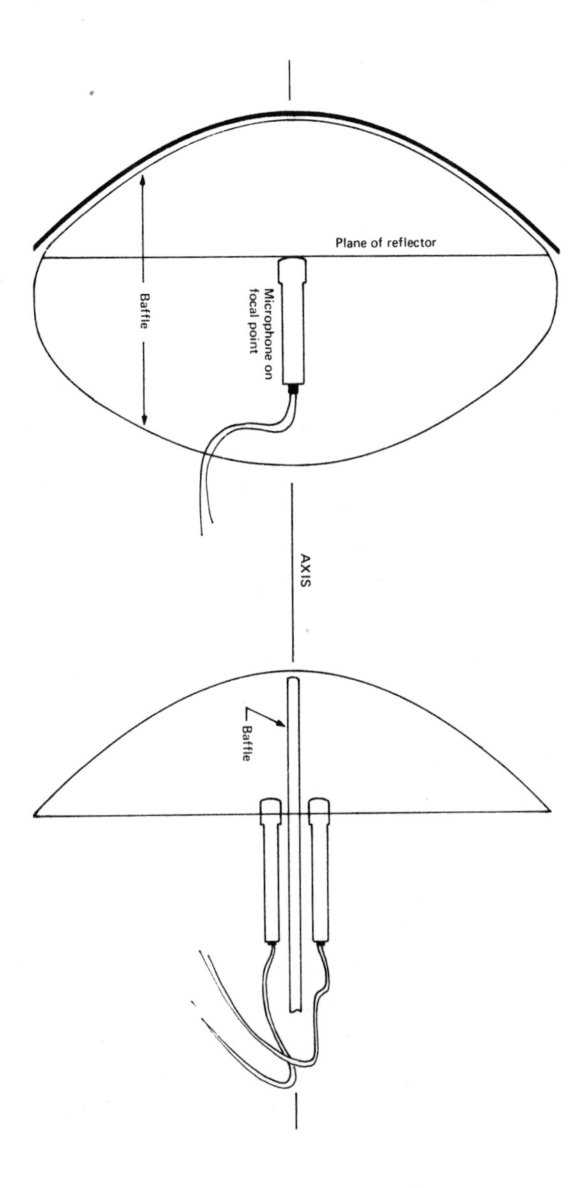

Fig. 11. Use of baffle to divide parabolic reflector for stereo-recording.

picture.

With two microphones being used together it is necessary for them to be in phase – if they are out of phase, then the signal from one will cancel that from the other. In stereo recording, out of phase microphones would result in the two separate tracks of the recording being out of phase so that, if they were played as one to produce mono, one would cancel the other; played as a stereo pair, the sounds from each speaker would be out of phase and, whilst a stereo picture would be produced, it would be difficult to locate the sounds within the picture.

Phasing of the system can be checked by placing the pair of microphones exactly side by side, facing the same way, and recording some speech. When the two tracks of the recording are then replayed together, as a mono, the result should be loud and clear with no loss of low frequency; if phasing is incorrect the result will be weak and distorted. Some recorders are provided with a mono switch to facilitate this procedure, but if the facility is lacking it is not difficult to add it in a line to a pair of headphones. Correction of the fault is achieved by inverting the phase of one microphone, this being done by reversing the connections of the two wires (usually red and black) in the microphone lead. When microphone extension leads are used they should be checked out to see that they do not invert the phase of one microphone due to incorrect wiring.

Phasing is important, of course, also in the reproducing system, and care must be taken to see that the leads to the speakers are correctly connected to the output sockets of the tape-recorder, or any external amplifier in use, to keep the speakers in phase. Any connecting lines between recorder and amplifier must also maintain correct phase.

Microphones used for stereo recording should be a 'matched' pair so as to give, as nearly as possible, the same overall frequency response, and the gain presented to each channel should be equal. Tape-recorders provide a separate gain control to each channel but the arrangement varies. In some cases the gain controls are side by side and two hands are required to operate them – not always easy in the field – and it is an advantage if they can be locked together so that they come up equally by turning only one; another arrangement provides an individual control to each channel with a master control operating on each.

When making stereo recordings, the approach is basically the same as if working in mono, but there are more points to watch. Microphones can, with care, be hand held, but even the smaller reflectors tend to be heavy for hand operation because of the extra microphone and the baffle; tripod mounting is really the answer. Once a recording has been commenced, the microphone set-up cannot be re-directed and the position the subject chooses to take up, within the sound picture, has to be accepted. If the microphones are near at hand, some improvement in the arrangement might be made before recording commences – so long as the subjects are reasonably static!

Much more patience is required and more tape run per minute of recording retained than is the case with mono. This means more listening, in the studio, to select the best sections, taking into consideration the balance and placing of any individual species and of any movement which takes place. If any editing is carried out, extra care is needed to check the relative position of a dominant species – it must not suddenly move from one spot to another because of an edit.

It should be clear, from what has been said, that if any copying, filtering or mixing of stereo recordings is to be carried out, it must be done with equipment which is capable of handling two or more separate channels at once. For copying, all that is required is two stereo tape-recorders. If the signal is to be filtered, or modified in any way, then the filters must be duplicated in order to handle each channel independently. When two stereo recordings are to be mixed, the mixer must be able to handle four separate channels – that is, two stereo pairs.

Programmes can be built up in stereo using methods described for mono work, but if track to track transfer is to be employed one tape-recorder will need to be a four channel (not quarter track) machine; such a machine is expensive and a four or six channel mixer would be the better answer.

There is nothing to prevent mono recordings being mixed into stereo ones, in fact some very effective results can be obtained by so doing. A clinical mono recording of an individual species can be added to a suitable stereo background and the balance between the two adjusted during the process.

Stereo mixers have what is known as a 'pan-pot' in each channel; this control determines to which channel, left or right, the signal is directed. The pan-pots of a stereo pair must be set to left and right to maintain a stereo recording, if both are set central, then the result will be mono. With a pair of channels taking a stereo recording, a third channel can be used to inject a mono recording, and the pan-pot on this channel used to 'place' the mono signal in the picture, anywhere from left to right; the fourth channel could be similarly used to add yet another mono recording. Such a process requires four tape-recorders; two mono to reproduce the mono recordings, a stereo one to reproduce the stereo recording, and another stereo one to re-record the mixed signals.

Cataloguing and Filing

Documentation

If tape recordings of natural sounds are to be used to their full value, then they must be accompanied by detailed documentation giving all the circumstances under which they were made and the equipment used. Rather than using a notebook, it is generally more convenient to dictate the details immediately after the recording has been made so that they are available on the tape when the recording is processed; notes on paper can then be confined to a list of the recordings made, during an expedition, for quick reference to spool contents.

The information which should be noted starts with the date, time and location; it should be followed by a good description of the habitat and notes of the species observed visually – others may well show up aurally when the recording is replayed. Weather conditions are also of sufficient importance to be recorded in some detail. These routine details having been attended to, attention should be given to the main subject of the recording by referring to its activity at the time; this is often the most difficult item because the recordist's attention will be principally centred on making a good recording, so a colleague's assistance as observer can be valuable. If the subject is a bird, the least that might be noted could be – 'singing from perch on conifer tree'; but why was he singing; did he have a nest nearby; was it territorial song; if it was an alarm, what was causing the alarm? There are many other points too which can be relevant; all should be noted.

The relevance of all this detail becomes apparent when the recording is to be compared with another of the same species and, particularly, if at any time it is to be offered for scientific research. When mixing recordings to make composite recordings or other programmes, the documentation helps in checking that

Serial No. 820B SPECIES COOT

Documentation Sheet

Rec No.	Qual Code	Voice & Behaviour	Weather Conditions, Remarks, etc	Location	Date	Time	Duration	Refl	Spd	Cut
1.	A	Calls of adult:	Calm – mild.	Alvecote Pools	22.2.76	07.40	3.10	–	15	L.F.A.
		good variety. Hoots		Nature Reserve						
		and lighter pitched	'squeaks'. Quite a lot of chasing,	Warwickshire.	815 gun to Nagra.					
		Good	movement and water sounds.							
2.	A.	A good 'splash-up'	Calm, sunny and warm		29.2.76	07.15	1.42	20/5	7½	–.
		with loud calling			AKG	D19C	– Sandberg			11.
		which gradually	dies down – territorial fighting.							

Fig. 12. Documentation Sheet

those selected will go together in a correct ecological manner.

All information relevant to any recording selected for inclusion in the library should be stored, against a reference number, on a suitable documentation sheet. A suitable layout is shown in Fig. 12. In addition to the details already discussed, provision is made for entering the duration of the piece filed; the diameter and focal length of any reflector used; the tape speed and any cut which has been, or should be, applied. Details of other equipment used are written across the columns.

Filing

One of the most aggravating things about tape recordings which have been stored away is not to be able to find quickly the one which is required, and yet a little care and attention on the matter of filing and indexing is all that is needed to produce an orderly library, making it possible to find one recording in a thousand within a minute.

The process of filing and indexing, and particularly of finding all the recordings of one species, is much simplified by arranging that all the recordings of that species are stored, one after the other, on one spool, a length of leader tape being inserted between each to indicate beginning and ending; a convenient arrangement is to use a colour coding. A coloured leader, say yellow, is used to indicate the beginning of a series of recordings of a species; on this leader will be printed, with black ball-point pen, the name of the species and its reference number from the indexing system. This yellow leader will be immediately followed by, say, an orange leader on which is printed the species reference number followed by . ./1, indicating that the tape that follows carries the first recording of the species; the second recording of the species will follow the first with only an orange leader between them and carrying species number followed by . . /2, and so on. The last recording of the series will be followed by a yellow leader indicating a change of species.

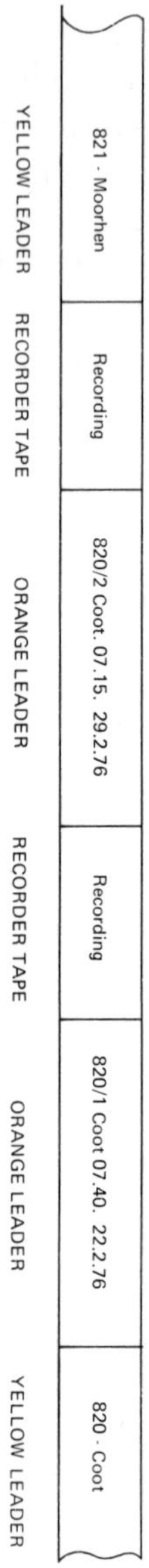

Fig. 13. Use of coloured leader tape to separate species and recordings.

When a recording is to be stored in such a system it is cut out of the spool on which it was made and, after the necessary leader has been added, spliced into the correct position on the storage spool. The spools are stored in boxes marked with the reference number they contain so that when a recording is required the spool carrying it is found very quickly. It is then necessary simply to count down a predetermined number of yellow and orange leaders, as the tape is spooled across the machine, to find the wanted one. The system requires that recordings are carried on one track only.

The stored recording should have a fade neither at the beginning nor at the end; in this way, when a copy is being made the recordist knows that he will go straight into the required signal as soon as the leader tape has passed the replay head, and that it will be there until the next leader appears – he can then arrange any fade required for the purpose in hand.

Indexing.

The idea that all the recordings of one species are stored together can be carried one step further in that related species, particularly in the case of birds, can be stored in groups; for instance, finches, tits and warblers. This requires that, if an index is being compiled as new species are recorded and entered, gaps in the numbering sequence should be left for them. One of the many pocket guide books can be of great assistance in building up a convenient index. Whatever form the index takes, the serial number is transferred to the record sheets which are filed alphabetically. The first column of this sheet provides space for the recording number relative to the species, so that the complete reference is – Serial No. / Recording No.

Outlets and Organisations

The B.B.C. Natural History Unit.

In 1948 the British Broadcasting Corporation purchased recordings of 204 species made by Dr. Ludwig Koch, the pioneer of wildlife sound recording, to form the basis of a very specialised section of the Corporation's Sound Archives. At the same time Dr. Koch rejoined the B.B.C. staff and continued to enlarge the collection until his retirement in 1951. At about this time, tape was beginning to take over from the discs which Koch had been using in the field, and with this new medium other workers greatly strengthened the collection of both British and Continental species; the work continues and by the mid '70s the library contained some 4000 cuts covering about 800 species. The latest development is the introduction of stereo recordings.

In 1957 the Natural History Unit was founded and based in Bristol, where it is still housed, with the main purpose of satisfying the B.B.C. requirements in broadcasting on radio and television; but its scientific interest has always been recognised. New recordings are continually being added by B.B.C. staff, but material is also purchased from freelance workers under a special agreement which allows the Corporation, whenever it wishes, to use the recording in any of its programmes anywhere in the world without further payment, but the recordist retains the copyright and may use it as he wishes. The main interest, of course, lies in recordings which are suitable for broadcast purposes, and habitat or 'atmosphere' recordings are included, the latter now particularly in stereo.

Anybody wishing to offer material should first write to the Sound Librarian, Natural History Unit, Broadcasting House, Whiteladies Road, Bristol BS8 2LR; details of the recording should be given and the tapes sent later, if requested.

The British Library of Wildlife Sound

The British Institute of Recorded Sound was established some years ago and is recognised as the National collection of sound recordings. In 1969 a new Department of the Institute was set up specifically to handle bio-acoustic recordings, and is known as the British Library of Wildlife Sounds; it already possesses a large collection including commercial, B.B.C., and privately made recordings which have been presented to it. The main aim of the library is to provide a National repository for natural history sound recordings which can be made available for scientific and cultural purposes. There are other collections of wildlife sound recordings in various parts of the world but B.L.O.W.S. is international in outlook; it obtains recordings from every continent and assumes special responsibility for British recordings and for recordings made in the Antarctic.

Recordings of all types of natural history sounds are required, and B.L.O.W.S. is only too willing to accept material from any amateur or professional recordist. A reasonably good technical quality is preferred, of course, but this does not rule out recordings made on inexpensive equipment; many recordings of doubtful technical quality can be of considerable scientific value, and contributors are invited to submit any recordings which they consider to be of interest. Whilst no payment is made, blank replacement tape is supplied and there is the great satisfaction that the work is held in a National Collection and that it is available to workers in other disciplines for scientific study.

There need be no fear about copyright for it remains with the recordist and is protected by very stringent rules. Users of the library have to sign an agreement not to infringe copyright in any way and that the recording is to be used for scientific purposes only; due acknowledgement must also be given in any scientific paper to the use of the recording contributed.

Full details can be obtained from the B.L.O.W.S. Librarian, The British Institute of Recorded Sound, 29, Exhibition Road,

London SW7.

The Wildlife Sound Recording Society

Some years ago a number of recordists, of which the author was one, recognised the need for an organisation which could band together all those who were actively engaged in making natural history sound recordings; the Society was formed at an inaugural meeting, attended by some eighteen interested persons, on March 30th 1968. The aim of the Society is to advise and assist in matters of technique and equipment, and to this end it publishes a Journal twice a year. In addition members are invited to submit examples of their work and demonstrations of recording techniques for inclusion in a tape-recorder programme which is circulated four times a year, this gives to members the opportunity of being able to compare their standards and techniques with those of others.

With a membership of over 250 it is probably the largest sound recording society in Britain, and there is no doubt that in its eight years of life The Society has already done much to improve the quality of wildlife recordings made in the British Isles, and to encourage those who are just starting. Many of the members are expert naturalists as well as recordists and so an opportunity is provided, in both Journal and Circulating Tape, for discussion on aspects of bio-acoustics. Because members are scattered throughout the British Isles the Society can be of assistance to other organisations in providing, or obtaining if not available, specialised recordings for research, broadcast or publication. So far it has not published any collection and, in order to protect members' copyright, the circulating tapes are strictly for the use of members only.

The widely distributed nature of membership makes it difficult to organise regular meetings convenient to all members; one is held in the spring and one in autumn, but these are supplemented by a certain number of more localised meetings arranged by members living close to one another. Full membership is available only to those residing in the British Isles but a form of Overseas membership – covering publications only, not the circulating tape – is available. Because it is a privately organised Society run by Honorary Officers only, no permanent address is given here but close contact is maintained with both the B.B.C. Natural History Unit in Bristol and the British Library of Wildlife Sounds in London.

The 'SCOTCH' Wildlife Sound Recording Contest.

The Contest, now going into its ninth year, is the only one known to deal exclusively with natural history subjects; it is sponsored and run by 3M United Kingdom Ltd., manufacturers of Scotch recording tapes. Divided into 'Open' and 'Novice' Categories it has classes for individual bird species; mammals, insects etc; and habitat/atmosphere. A special class for stereo recordings has recently been added.

The Contest has done a great deal to encourage wildlife sound recording in Britain and has brought to light a number of unusual and important recordings.

The British Amateur Tape Recording Contest.

The long established British Amateur Tape Recording Contest now has a special class dealing with 'Sounds from nature'. In addition, prepared programmes having nature as their subject are suitable for entry in other classes – documentary, for instance. The Contest, which does a lot to encourage creative tape-recording, is run by a voluntary organising committee and sponsored by various manufacturers. A selection of the tapes entered in this Contest are forwarded for entry in the International Amateur Tape Recording Contest.

Copyright

Good recordings can have a commercial value, for they are

frequently sought by film and recording companies, and for broadcasting purposes. It is very difficult indeed to set any value on a recording, the fees offered by the larger companies are invariably fair but in other approaches it might be prudent to exercise care and, if necessary, take advice. The most important thing is *never* to sell all rights in the recording – that is, the copyright. In the case of recording companies the most that should be allowed is the reproduction right in the context of one specified title only and, in the case of broadcasting, world non-exclusive rights. This arrangement means that the copyright of the recording remains with the author who then retains the right to sell it however many times it comes into demand.

Odd Ends

When using two tracks on a spool of tape, always use the same 'take-up' spool and have it clearly marked side 1 and side 2 – it is a good insurance against using the same track twice.

When one track only is used, be sure the tape has a red *trailer* – this is a warning not to use it again.

A white leader is a considerable help in threading a tape in the dark.

If small spools are loaded from large ones (an economical way to purchase), always attach leader tape at each end; this prevents the tape getting 'dented' by the slots in the spool hub. If tape so dented gets into the middle of a spool, by splicing together partly empty spools, the dents will produce a rhythmic bumping in the recording.

When cutting out recordings for filing away in a library, be quite sure to make perfect splices when re-joining the tape which is to go back into stock. Failure to do so will result in noise when the splice goes over the recording head. Do not use tape containing many splices for work in the field.

A self-adhesive label, app. ½ x 2 inches, on each side of a spool is useful for adding a reference number and brief notes when in the field.

As far as possible, standardise the connectors on all microphone leads. The fact that different microphones and tape-recorders use different plugs and sockets is easily overcome by making up a short adaptor lead for each.

If a stereo recorder is being used to make mono recordings, wire the microphone plug to put the signal on to both tracks; this gives nearly the equivalent of full track recording and will improve quality.

If working in stereo, colour code leads for left and right channels.

A selector box is useful when several microphones are out

at once but only one can be used at a time; it saves having to change plugs. The box has several input sockets into which the microphone leads are plugged, and one output lead which is plugged into the tape-recorder. The required microphone is then selected by a switch in the box.

When a microphone is placed on a stand some distance from the recordist, it is very aggravating if a bird selects a 'song-post' which is just 'off-mike'; a form of remote control would be useful. Small 12 volt d.c. motors with a gear box attached can be found at surplus stores and it is not a difficult matter to make the necessary modifications to fit them between the tripod head and the microphone. It is, of course, necessary to have a 12 volt supply which, unless it can be obtained from the tape-recorder supply, will require carrying an extra battery. An extra lead will also be necessary to carry the current to the motor unless arrangements can be made to use the screen and one line of the microphone lead – without damaging the microphone or the tape-recorder. Reversal in the polarity of the line, by changing the connections to the battery or inserting a switch for the purpose, can be used to cause the microphone to turn to left or right.

When purchasing batteries it is advisable to take one more cell than is required to load the recorder; this is a safeguard against one of the cells being a 'dud'. One dud cell in a set will result in failure of the machine and the test with a light bulb is inadequate – it is only when the recorder motor puts a load on that the dud cell will fail.

Rechargeable cells are available for most tape-recorders – check whether such a cell is the type which should always be kept fully charged. When on a holiday which restricts access to a mains supply such a cell will be of no use unless it can be recharged from the car battery.

Clothes worn in the field are important. Obviously, at times, they must provide protection and warmth but what may not be so obvious is that they must be *quiet*. Some materials, nylon in particular, are very noisy and produce a high frequency 'swish' at the slightest movement – a noise that can so easily ruin a recording.

Finally, to reiterate what has been said before, little progress will be made in this fascinating branch of tape-recording without a very liberal portion of PATIENCE.